PHYSICS PRACTICAL SCHEME OF WORK

FOR USE WITH THE IB DIPLOMA PROGRAMME
First Assessment 2016

Michael J. Dickinson

For:
Mum and Dad

ACKNOWLEDGMENTS

© Michael J. Dickinson 2014
dickinson.physics@gmail.com

First edition published 2012
Second edition published 2014

ISBN-10:   1494987899
ISBN-13:   978 - 1494987893

This work has been developed independently from and
is not endorsed by the International Baccalaureate (IB)

Front cover:
Photograph: "Soap Bubble Sky", by Brocken Inaglory
Graphic design: Michael J. Dickinson

Thanks to:
Silvio Augustin, for his work that was used
as the Individual Investigation example.

All my IB students between 1997 and 2009 who acted as guinea-pigs
and proof-readers as I wrote and/or modified each lab sheet.

Also, any student and teacher out there that took the time to
contact me in order to point out typos and mistakes in the first
edition of this practical guide – your feedback has helped to make
this second edition more accurate and reliable than the first.

Copyright:
Every effort has been made to trace the holders
of copyright and acknowledge them. If any have
been overlooked please contact me so that the
necessary adjustments can be made for the next
edition of this book.

PHYSICS PRACTICAL SCHEME OF WORK
FOR USE WITH THE IB DIPLOMA PROGRAMME
First Assessment 2016

CONTENTS

PART 1

Preface to the 1st Edition	iv
Preface to the 2nd Edition	v
The Practical Scheme of Work (PSOW)	vii
Distribution of Time	ix
Syllabus Coverage	ix
Planning the PSOW	ix
Types of Practical Experiences	x
Use of Information and Communication Technology (ICT)	x
Assessment	x
The Group 4 Project	xi
Form 4/PSOW	xii
Form 4/CSS	xii
Form 4/IA	xiii
Sample Form 4/PSOW (Practical Scheme of Work)	xiv
Sample Form 4/CSS (Individual Candidate Cover Sheet)	xvi
Sample Form 4/IA (Internal Assessment Checklist)	xvii

PART 2

The Investigation	1
Assessment of the Investigation	3
Weighting of the Criteria	3
Acceptable Levels of Teacher Assistance	4
Internal Assessment Rubrics	5
Personal Engagement	*5*
Exploration	*5*
Analysis	*6*
Evaluation	*6*
Communication	*7*
Sample Investigation	8
Marking the Sample Investigation	18

PART 3 | REQUIRED EXPERIMENTS 21

Determining the Acceleration of Free-Fall 25
Determining Specific Heat Capacity by the Electrical Method 26
Determining Specific Heat Capacity by the Method of Mixtures...................27
Investigating Specific Latent Heat of Vaporization of Water.........................28
Investigating Specific Latent Heat of Fusion of Ice 29
Investigating Boyle's Law 30
Investigating Charles' Law and Absolute Zero 31
Investigating Lussac's Law (The Pressure Law) 32
Investigating Resonance – Determining the Speed of Sound 33
Determining the Refractive Index of Glass by Real and Apparent Depth 34
Investigating Refraction of Light, Refractive Index and Critical Angle 35
Investigating Factors that Affect Resistance..36
Determining the EMF and Internal Resistance of a Cell 37
Investigating Radioactive Decay and Half-Life (Simulation using Dice) 38
Determining the Wavelength of Laser Light using Young's Double Slits 39
Investigating a Diode Bridge Rectification Circuit.....................................40
Investigating the Compound Microscope 41
Investigating a Simple Optical Astronomical Refracting Telescope 42

PART 4 | OTHER SUGGESTED PRACTICAL ACTIVITIES...43

Investigating Murphy's Law 45
Investigating Errors and Uncertainties in Experiments 46
Investigating Uncertainties – Measuring Instrument Circus.........................47
Determining Stiffness of Steel by the Oscillations of a Hacksaw Blade 48
Investigating the Vibrations of a Loaded Metre Rule 49
Investigating the Range of Gamma Radiation in Air...................................50
Investigating Forces in Equilibrium..51
Investigating the Fall of a Coffee Filter 52
Investigating Parabolic Motion..53
Investigating the Flight of an Elastic Band 54
Investigating the Simple Pendulum 55
Investigating the Stopping Distance of a Bicycle 56
Investigating the Torsional Pendulum 57
Investigating Newton's 2nd Law of Motion using a Ticker Timer....................58
Investigating Hooke's Law 59
Investigating Springs 60
Investigating Work Done and Energy Transferred on an Inclined Plane 61

Investigating the Ballistic Pendulum 62
Investigating the Power and Temperature of the Sun 63
Investigating Rate of Cooling 64
Determining the Temperature of a Wire by Expansivity 65
Investigating Ping-Pong Balls 66
Investigating the Pressure / Volume Relationship for a Balloon 67
Investigating Malus' Law 68
Investigating Brewster's Law 69
Determining the Wavelength of Light using a Diffraction Grating 70
Investigating Melde's Experiment 71
Investigating the Power of an Electric Heater 72
Investigating Resistors in Series and Parallel 73
Verifying the Equation "F = B I L" Using a Current Balance 74
Investigating Electromagnets 75
Investigating Circular Motion 76
Determining Energy Density of Fuels 77
Investigating Energy 78
Investigating the Superposition of Sound Waves 79
Investigating Magnets 80
Investigating Lenz's Law using the Motion of a Falling Magnet 81
Investigating Electromagnetic Induction 82
Investigating the Efficiency of a Transformer 83
Investigating Energy Transfer and Energy Loss of a Rolling Ball 84
Investigating the Focal Length of a Converging Lens 85

PART 5 THE GROUP 4 PROJECT 87
Science IB Group 4 Project – Introduction 90
Science IB Group 4 Project – Questions and Answers 91
Science IB Group 4 Project – Planning 92
Science IB Group 4 Project – Evaluation 94
Science IB Group 4 Project – Summary 95

PART 6 APPENDIX 97
Possible Topics for Further Investigation 99

PREFACE TO THE 1ST EDITION

My first exposure to the International Baccalaureate program was back in about 1995, when I moved overseas to teach at the Colegio Anglo Colombiano, Bogotá, Colombia. Before I left the UK though, while visiting a school in Liverpool that ran the IB Diploma, the head of science there tried to instill in me just how important it was to understand the requirements of the Internally Assessed (IA) portion of the syllabus. He offered me all sorts of advice regarding the experiment side of IB Sciences and I left that day with my first Practical Scheme of Work (PSOW) for the IB Diploma Physics course.

In the first 6 months of teaching at "The Anglo" in Bogotá, I was sent to Cali, in the South of Colombia, on an IB Physics workshop – it was a level 1 workshop for rookies just like me and a large portion of the course was dedicated to IA practicals. It was run by Brian Seve, who was at the time the Head of Science at The English School in Bogotá. I remember vividly a large science lab set up with a dozen or so experiments around the edge. Each experiment had an accompanying student help sheet and I spent a couple of afternoons working my way around the experiments so that I would have a larger repertoire of labs to perform with my students when I returned to Bogotá. I took Brian's 4/PSOW away with me which I used extensively for the following few years.

A couple of years into my Anglo contract, my first cohort of seniors sat their IB exams. They performed reasonably well on the externally examined part, but I remember feeling disappointed with their results for the internally assessed work. When the marks had been published, together with the examiner's comments, I found to my dismay that all my students had been marked down! It transpired that I had been giving "too much information" to my students. For example, if I had told them both the dependent and independent variables and then assessed the lab on the Design (D) criteria students could not receive a "complete", even though I had marked it as such. Similarly, if I had given students the outline of a results table and suggested to them which graph they should draw, students could not receive a "complete" for Data Collection and Processing (DCP). A similar thing happened to the third criteria, Conclusion and Evaluation (CE) if I had given too much guidance.

It wasn't until I had moved to Helsinki, Finland in 1999 that I finally felt that I understood how much (or how little) information was acceptable for these student lab help sheets. Another colleague, Trevor Wilson, who had been a Workshop Leader for Physics for a number of years, critiqued each of the lab sheets in my PSOW noting which criteria we could assess if we used them. There was absolutely nothing wrong with Brian's lab sheets; it was just that I should have been a little more discerning about which ones were suitable to assess which criteria.

A few years later I became an IB Moderator for Physics IAs and I remember receiving a sample of student work and feeling very nervous about the grade that I was about to award. The teacher had awarded one of his students 47/48, but I noticed that the teacher had told the students both the independent and dependent variables, when he was assessing the "Design" criterion. For the experiments that were being assessed for the "Data Collection and Processing" criterion, the teacher had told his students what data to collect, had given students results tables with the column headings already complete (so that all they had to do was record their data) and had told them what graph to draw and how to analyze it. The conclusions and evaluations of the labs were in the form of answers to a series of questions that the teacher had provided. I moderated the work down to 3/48, as the sample was not a true assessment of the capability of the student, but was more an indication of how well the student could follow instructions that the teacher had provided.

It should be noted that if the moderator finds a teacher's marking accurate, then no moderation factor is assigned to that teacher. However, if the moderator feels that the teacher has been a little too generous in their marking, then a moderation factor is applied to that teacher such that all the students in that teacher's classes could possibly be marked down. Of course, if the teacher has been too harsh in their marking then the moderation factor applied to the teacher will benefit the students by increasing their overall IA mark.

I hope that you will find the information contained within this PSOW guide helpful and I wish you and your students every success in the internally assessed part of the IB Physics course.

Mike Dickinson – March 2012

Preface to the 2ⁿᵈ Edition

Building on the success and, for the most part, the very positive feedback I received from the first edition of the Physics Practical Scheme of Work, I found myself drawn to producing a second edition, in the hope that some of the more radical changes that the IBO have introduced for this latest incarnation of the subject guide can be addressed in an understandable way. Hopefully, teachers might more swiftly embrace these changes, resulting in a quicker modification of their previous practices, to the benefit their students.

In October 2013 I completed online "upskilling" training, during which a number of Physics Workshop Leaders (WSLs) – about 20 in my group – were given an opportunity to take a sneak peak at an early draft of the latest syllabus. The intention of the training was to allow WSLs to deliver workshops to physics teachers starting in the first half 2014, 6-7 months prior to the commencement of the teaching of the latest syllabus. The training produced some lively debate and sharing of ideas, opinions and resources that I found very useful in my understanding of the changes to the new syllabus.

Some of the requirements have hung over from the incumbent syllabus, while others have been introduced and will require some explanation. For example, the Group 4 Project has remained largely the same as in the previous guide; as far as I can see, the only difference in the Group 4 Project is that it is no longer assessed whereas the previous syllabus used the Personal Skills criterion to assess the students' participation. The setup and collaborative, cross-discipline nature of the Group 4 has remained intact as has the 10 hour duration of the project. This enables schools to continue to use existing models that have worked for them in the past without the necessity of changing this aspect of the PSOW.

The overall timing of the Practical Scheme of Work has also remained identical to the timing that was defined in previous subject guides in that Standard Level students will be expected to have carried out approximately 40 hours of practical work over the course of the two year program. The number of hours that Higher Level students are expected to have completed on practical work remains at 60 hours

One change in the structure of the Practical Scheme of Work is in the way that these 40 hours and 60 hours are divided up. In the cases of both Standard and Higher levels, 10 hours of their total time are allocated to the Group 4 Project. However, a new requirement introduces a 10 hour allocation to "The Investigation". The Investigation will constitute the only piece of Internally Assessed (IA) work that will be sent to the IBO for moderation, and so a complete understanding of the requirements of The Investigation is imperative to ensure optimal success for our students.

Another major change is in the design of new rubrics that are to be used to assess the students' completion of The Investigation. The five criteria, Design (D), Data Collection and Processing (DCP), Conclusion and Evaluation (CE), Personal Skills (PS) and Manipulative Skills (MS) have given way to five new ones; Personal Engagement (PE), Exploration (EX), Analysis (A), Evaluation (EV) and Communication (C). It will be interesting to see how these new criteria align with the old ones, but there does seem to be some overlap. For example, there are similarities between Design and Exploration, between Data Collection and Processing and Analysis and between Conclusion and Evaluation and Evaluation. Similarly, the Personal Skills criterion has given way to Personal Engagement and Manipulative Skills has been discarded in favor of Communication. Another change is that the "complete", "partial", "not at all" (c,p,n) notation has been removed and in its place are a new set of rubrics whose structure is aligned with that used for the Extended Essay.

Finally, the remaining hours (after the 10 each for the Group 4 Project and the Investigation) are given over to general laboratory investigations and practical experiences. A slight difference here is that there are some investigations that are compulsory as they have the potential to be tested in the external assessed component (examinations). I have of course included some guidance and suggestions on these compulsory investigations within this book.

Mike Dickinson – March 2014

THE PRACTICAL SCHEME OF WORK (PSOW)

The Practical Scheme of Work (PSOW)

Part of the requirement of the IB Syllabus for any of the experimental sciences in Group 4, is the completion of a "Practical Scheme of Work" or PSOW. This course of experimental study forms a very large part of a teacher's commitment and responsibility to ensure that students have been well prepared for experimental scientific endeavor, should they choose to pursue this type of career for their future. It is a responsibility since 20% of a student's overall grade is gained depending on how well they perform on "The Investigation", an in-depth personal project which forms a large part of the practical scheme of work. The teacher must be committed since there is a significant time consideration to be made when designing the learning plan for the two year course.

Distribution of Time

Standard Level students must complete a minimum of 40 hours of practical work, while Higher Level students must complete at least 60 hours (this time includes a maximum of 10 hours spent on the Group 4 Project but excludes and time spent writing up lab work).

	SL hours	HL hours
Practical activities	*20*	*40*
Individual investigation (internal assessment)	*10*	*10*
Group 4 project	*10*	*10*
Practical scheme of work	**40**	**60**

Students in Standard and Higher Level may perform some of the same investigations – this is particularly useful when, in smaller schools, SL and HL classes are often combined. Only 2 to 3 hours of investigative work can be carried out after the deadline for submitting work to the moderator and still be counted in the total number of hours for the practical scheme of work.

Syllabus Coverage

When designing the Practical Scheme of Work, teachers should consider both breadth and depth of the syllabus being covered. It may be tempting, for example, to perform many experiments which cover the Mechanics and Electricity topics at the expense of covering the sections within, say, the Quantum and Nuclear Physics topic. This is only natural as some topics lend themselves to a wider range of experimental study than other, but the teacher should attempt to spread the investigations over as many topics as possible. However, it is NOT mandated that students should carry out investigations from every syllabus topic.

Planning the PSOW

Teachers are free to formulate their own practical schemes of work by choosing practical activities according to the requirements outlined. Their choices should be based on:
- subjects, levels and options taught
- the needs of their students
- available resources
- teaching styles

Each scheme must include some complex experiments that make greater conceptual demands on students. A scheme made up entirely of simple experiments, such as ticking boxes or exercises involving filling in tables, will not provide an adequate range of experience for students.

Strict adherence by the teacher to the IBO's Internal Assessment guidelines is important so that each student fulfills their obligations in terms of both time and content requirements. That said, within the PSOW, The Investigation enables students to demonstrate the application of their skills and knowledge, and to pursue their personal interests, without the time limitations and other constraints that are associated with written examinations, and so in this regard, can be made into one of the more enjoyable aspects of the Physics course.

Types of Practical Experiences
The practical programme is flexible enough to allow a wide variety of practical activities to be carried out. These could include:
- short labs or projects extending over several weeks
- computer simulations
- using databases for secondary data
- developing and using models
- data-gathering exercises such as questionnaires, user trials and surveys
- data-analysis exercises
- fieldwork

[Physics Diploma Programme Guide, 2007. ©IBO]

Although the requirements for Internal Assessment are centered on The Investigation, the different types of practical activities that a student may engage in serve other purposes, including:
- illustrating, teaching and reinforcing theoretical concepts
- developing an appreciation of the essential hands-on nature of much scientific work
- developing an appreciation of scientists' use of secondary data from databases
- developing an appreciation of scientists' use of modeling
- developing an appreciation of the benefits and limitations of scientific methodology.

[Physics Diploma Programme Guide, 2014. ©IBO]

Use of Information and Communication Technology (ICT)
One of the aims of IB is to "develop and apply the students' information and communication technology skills in the study of science". The use of information and communication technology (ICT) is encouraged in practical work throughout the course, whether the investigations are assessed using the IA criteria or otherwise.

> Aim 6. Develop experimental and investigative scientific skills including the use of current technologies.
> Aim 7. Develop and apply 21st-century communication skills in the study of science.

It is not necessary to use ICT in every piece of practical work but, in order to carry out aims 6 & 7 in practice, students will be required to use each of the following software applications at least once during the course.
- Data logging in an experiment
- Software for graph plotting
- A spreadsheet for data processing
- A database
- Computer modeling / simulation

[Physics Diploma Programme Guides, 2007 & 2014. ©IBO]

Assessment
Not every piece of practical work has to be assessed in accordance with the IB Internal Assessment criteria. There are often times when students use pieces of apparatus to support their learning and yet, the work is not going to be assessed. The hours spent on informal, non-assessed practical work can also be included in the 40 or 60 hours. In fact, the only piece of student work that must be assessed is The Investigation. However, I feel that it is advisable to use the marking criteria rubrics formatively for at least some of the other practical activities so that students get a feel for how they are to be assessed before the summative Investigation is undertaken and sent to IBO for moderation.

A minimum number of investigations is not specified by IBO, so teachers are quite at liberty to perform many smaller experiments or fewer, more in depth investigations, as they see fit. My own preference is to do a mixture of the two – in this way students can undertake labs which can be assessed used all 5 of the assessment criteria before they attempt "The Investigation".

I have included some experiments that can be completed within the duration of a 60 - 80 minute lesson and others which might take multiple lessons to complete. The longer, multiple-lesson practicals will allow students the opportunity to explore a scientific topic with much more depth and so might be assessed using the Exploration (EX) criterion – although I realize that this is not perfect as doling out of the topic itself removes the opportunity for students to demonstrate a "personal significance, interest or curiosity". At the same time, I have included many examples of experiments that can be used to assess a student's ability to collect, manipulate and present data, so fulfilling the IB requirement of the Analysis (A) criterion of the PSOW. The third assessment criterion is Evaluation (EV) and again, there are numerous opportunities within the included guide to assess this. I have included a suggestion (in the form of an assessment marking grid) on the top-right corner of each of the lab sheets contained within this book for which assessment criteria that particular lab might be appropriate. Please use my suggestions only as a guide.

Most of the lab sheets contained within this book can also be used to assess a student's "Personal Engagement", and of course all the practicals can be used to assess students' "Communication" skills; the last of the 5 assessment criteria. Some of them contain apparatus which can be quite tricky to set up and manipulate in order to obtain good data – the Current Balance for example is particularly difficult to assemble, but once working the students can obtain some great data to verify the equation.

The Group 4 Project
One investigation that MUST be completed by ALL students is the Group 4 Project. The Group 4 Project is an opportunity for students across the range of scientific disciplines to work together in order to investigate a topic from a range of angles. This interdisciplinary project allows Chemistry, Biology and Physics students (together with Computer Science, Sports Exercise & Health Science and Design Technology – if offered at your school) to engage in a collaborative project. Students taking Environmental Systems & Societies are not required to participate in the Group 4 Project. The emphasis of this project is on process rather than product and so, the Group 4 Project is not assessed.

At the back of this book, I have included an example of the Group 4 Project that I undertook a few years ago while teaching Physics at the United Nations International School (UNIS), in Hanoi. The overall theme of the project was "Science on and around the School Campus" – students investigated a diverse range of sub-topics within this general theme. A variety of science disciplines was represented in each of the small groups and while each student within the group was responsible for their particular area of science, the whole group collaborated together to make these different aspects of science to meld together into a coherent final report.

Physics Practical Scheme of Work – For use with the IB Diploma Programme – First Assessment 2016

Form 4/PSOW

A full list of experimental work that candidates have been exposed to over the two year course of study should be recorded on the 4/PSOW form. The form also contains the details regarding the time spent on each practical activity and the ICT component(s) in which the candidates' demonstrated their competence.

The following MUST be included on the form:

At the top of the form:
- Submit to   Moderator
- Arrival date   April / October (depending on the session of the exam)
- Session   May / November
- School number   00 XXXX
- School name   Use words rather than abbreviations
- Subject   Physics
- Level   Standard / Higher level

In the body of the form:
- Date(s)   The dates that the practical work was carried out
- Outline of experiment   This doesn't have to be too detailed – often the title is sufficient
- ICT   Number 1 – 5 corresponding to the ICT type
- Topic / option   Topic and sub-topic from the syllabus guide
- Time (hrs)   I normally use 0.5 hours as a minimum
- Group 4 Project   The title of the Group 4 Project is included

At the end of the form:
- Teacher's name, Teacher's signature and date.

The IB provide the 4/PSOW form in section 4 of the "Handbook of procedures for the Diploma Programme" – this used to be called the "Vade Mecum" until IB realized that nobody understood what "Vade Mecum" meant. On the following pages, I have provided my own version of this form as an example of how it might be filled out.

Form 4/CSS

The 4/CSS form is new for this latest version of the IA sample submission process. Teachers used to include all the individual student assessment information on the 4/PSOW form; an individualized copy of which was included as the cover page for each student's IA moderation sample. In order to streamline the flow of information, only one 4/PSOW form will be included for the entire moderation sample, but an individual 4/CSS form will be included for each candidate. The mark out 24 for each student's individual Investigation should be recorded on the Form 4/CSS, together with the following information.

At the top of the form:
- Submit to   Moderator
- Arrival date   April / October (depending on the session of the exam)
- Session   May / November
- School number   00 XXXX
- School name   Use words rather than abbreviations
- Subject   Physics
- Level   Standard / Higher level
- Candidate name   Use the name provided to IB when the student registered
- Candidate number   00 XXXX XXX

In the body of the form:
- Date — The date that The Investigations was carried out
- Time (hrs) — I normally use 0.5 hours as a minimum
- Criteria mark — The mark for each of the 5 individual criteria are included
- Total mark — Total mark out of 24 for the candidate is included
- Hours — Total time participation in the PSOW

At the end of the form:
- Teacher's name, Teacher's signature and date.
- Candidate's name, candidate's signature and date.
- Candidate declaration regarding participation in the Group 4 Project and academic honesty.

It is important that teachers retain a copy of the 4/CSS form using the Save As function or by printing a copy. The form should be completed in the working language of your school (English, French or Spanish). After completing the form, it must be printed and signed by the teacher and candidate to confirm the authenticity of the work.

Form 4/IA

In addition to the 4/PSOW form (one form for each school) and the 4/CSS form (one for each candidate), teachers are encouraged to complete the 4/IA form. The 4/IA form is a checklist to ensure that teachers have completed all the necessary requirements for the IA moderation sample (normally a sample consisting of the Investigation for 10 students). The checklist itself should not be submitted to IBO with the moderation sample.

- I have read section XXXX and XXXX and section XXXX in the handbook. (IBO's assessment department will enter the section numbers when known).
- Internal standardization has taken place where two or more teachers are responsible for the internal assessment of candidates.
- A form 4/PSOW is included, signed and dated by the teacher.
- Photocopied material is legible (ideally, original work should be sent to the moderator).
- The criteria Personal Engagement (PE), Exploration (EX), Analysis (A), Evaluation (EV) and Communication (C) have all been assessed.
- The levels marked for each of the criteria, PE, EX, A, EV and C have been clearly entered on each candidate's Individual Candidate Cover Sheet (form 4/CSS).
- The Investigation and accompanying teacher instruction sheets for each candidate in the sample set are enclosed.
- The title of the Group 4 Project is included in the outline of experiments in the 4/PSOW.
- Each candidate has written a reflective statement on his/her involvement in the Group 4 Project on the 4/CSS form
- The experiments/dates on which the students experienced specific ICT applications have been flagged.
- Each candidate has signed a declaration on academic honesty.
- The final mark out of 24 for internal assessment has been recorded on each candidate's Individual Candidate Cover Sheet (form 4/CSS) and has been entered on the internal assessment option on IBIS.
- Moderation samples for atypical candidates have been identified and the work of another candidate with the same or similar mark in addition to the candidate's work has been included.

Physics Practical Scheme of Work – For use with the IB Diploma Programme – First Assessment 2016

SAMPLE FORM 4/PSOW
(PRACTICAL SCHEME OF WORK)

> A guess at what the 4/PSOW Form might look like has been made for this publication. The actual form will be available in the IB Handbook of Procedures at the start of the 2015-16 school year.

Submit to: **Moderator**   Arrival date: **20 April 2016**   Session: **May 2016**

School code: **00001**   School name: **International School of South East Asia**

Subject: **Physics**   Level: **Higher Level**

Please fill in the ICT column using the numbers below to show when students experienced each of these applications:
1– Data logging,   2 – Graph plotting software,   3 – Spreadsheet,   4 – Database,   5 – Computer model / simulation

Date(s)	Outline of experiments / investigations / projects	ICT	Topic / option	Time (hrs)
	Investigating Murphy's Law	----	1.1	3.0
	Investigating Errors and Uncertainties in Experiments	----	1.2	1.0
	Investigating Uncertainties – Measuring Instrument Circus	----	1.2	1.0
	Determining Stiffness of Steel by the Oscillations of a Hacksaw Blade	(1),2,3	1.2	1.0
	Investigating the Vibrations of a Loaded Meter Rule	2,3,4	1.2	1.0
	Investigating the Range of Gamma Radiation in Air	2,3,4	1.2	1.0
	Investigating Forces in Equilibrium	----	1.3	1.0
	Determining the Acceleration of Free-Fall*	(1),2,3	2.1	1.0
	Investigating the Fall of a Coffee Filter	2,3	2.1	3.0
	Investigating Parabolic Motion	(1),2,3	2.1	1.0
	Investigating the Flight of an Elastic Band	2,3,5	2.1	3.0
	Investigating the Simple Pendulum	2,3	2.2	3.0
	Investigating the Stopping Distance of a Bicycle	2,3	2.2	3.0
	Investigating the Torsional Pendulum	(1),2,3	2.2	3.0
	Investigating Newton's 2nd Law of Motion using a Ticker Timer	(1),2,3	2.2	1.0
	Investigating Hooke's Law	2,3	2.2	1.0
	Investigating Springs	4,5	2.2	3.0
	Investigating Work Done and Energy Transferred on an Inclined Plane	2,3	2.2	1.0
	Investigating the Ballistic Pendulum	----	2.4	0.5
	Determining Specific Heat Capacity by the Electrical Method*	1,2,3	3.1	1.0
	Determining Specific Heat Capacity by the Method of Mixtures*	----	3.1	0.5
	Investigating Specific Latent Heat of Vaporization of Water*	2,3	3.1	1.0
	Investigating Specific Latent Heat of Fusion of Ice*	----	3.1	0.5
	Investigating the Power and Temperature of the Sun	(1),2,3	3.1	1.0
	Investigating Rate of Cooling	1,2,3,5	3.1	1.0
	Determining the Temperature of a Wire by Expansivity	----	3.1	0.5

	Experiment			
	Investigating Boyle's Law*	1,2,3	3.2	1.0
	Investigating Charles' Law and Absolute Zero*	2,3	3.2	1.0
	Investigating Lussac's Law (The Pressure Law)*	(1),2,3	3.2	1.0
	Investigating Ping-Pong Balls	(1),2,3	3.2	3.0
	Investigating the Pressure / Volume Relationship for a Balloon	(1),2,3	3.2	1.5
	Investigating Resonance – Determining the Speed of Sound*	2,3	4.2	1.0
	Investigating Malus' Law	1,2,3	4.3	1.0
	Investigating Brewster's Law	1,2,3,(4)	4.3	1.0
	Determining the Refractive Index of Glass by Real and Apparent Depth*	----	4.4	0.5
	Investigating Refraction of Light, Refractive Index and Critical Angle*	2,3	4.4	1.0
	Determining the Wavelength of Light using a Diffraction Grating	4,5	4.4	1.0
	Investigating Melde's Experiment	2,3	4.5	1.0
	Investigating Factors that Affect Resistance*	----	5.2	3.0
	Investigating the Power of an Electric Heater	2,3	5.2	1.0
	Investigating Resistors in Series and Parallel	----	5.2	3.0
	Determining the EMF and Internal Resistance of a Cell*	4	5.3	1.0
	Verifying the Equation "F = B I L" Using a Current Balance	2,3	5.4	1.5
	Investigating Electromagnets	2,3	5.4	3.0
	Investigating Circular Motion	2,3	6.1	1.0
	Investigating Radioactive Decay and Half-Life (Simulation using Dice)*	2,3,5	7.1	1.0
	Determining Energy Density of Fuels	4	8.1	1.0
	Investigating Energy	1,2,3,4,5	8.1	5.0
	Determining the Wavelength of Laser Light using Young's Double Slits*	2,3,4	9.3	1.0
	Investigating the Superposition of Sound Waves	----	9.3	1.0
	Investigating Magnets	----	10.1	3.0
	Investigating Lenz's Law using the Motion of a Falling Magnet	1,2,3	11.1	1.0
	Investigating Electromagnetic Induction	2,3	11.1	3.0
	Investigating a Diode Bridge Rectification Circuit*	1,2,3	11.2	1.0
	Investigating the Efficiency of a Transformer	2,3	11.2	1.0
	Investigating Energy Transfer and Energy Loss of a Rolling Ball	----	B.1	1.0
	Investigating the Focal Length of a Converging Lens	2,3	C.1	1.0
	Investigating the Optical Compound Microscope*	2,3	C.2	1.0
	Investigating a Simple Optical Astronomical Refracting Telescope*	----	C.2	1.0

* Required experiments

	Group 4 Project – Investigating energy efficiency of systems on the UNIS campus	1,2,3,4,5	----	10.0
	The Investigation	1,2,3,4,5	----	10.0

To be completed by the teacher:

Name: **Michael John Dickinson**   Signature: **M. J. Dickinson**   Date: **March 5th 2016**

Physics Practical Scheme of Work – For use with the IB Diploma Programme – First Assessment 2016

Sample Form 4/CSS
(Individual Candidate Cover Sheet)

> A guess at what the 4/CSS Form might look like has been made for this publication. The actual form will be available in the IB Handbook of Procedures at the start of the 2015-16 school year.

Submit to: **Moderator**   Arrival date: **20 April 2016**   Session: **May 2016**

School code: **00001**   School name: **International School of South East Asia**

Subject: **Physics**   Level: **Higher Level**

Candidate name: **Bond, James**   Candidate number: **00001 007**

Date	Outline of experiments / investigations / projects (include title and a brief description)	Levels awarded by Teacher					Time (hrs)
		PE	EX	A	EV	C	
	The Investigation	1/2	4/6	5/6	4/6	3/4	10.0

Please retain a copy of this form using the *Save As* function or by printing a copy.
Complete this form in the working language of your school (English, French or Spanish)
After completing this form, it must be printed and signed by the teacher and candidate to confirm the authenticity of the work.

Total Mark: **17**/24

Total Time: **62**/hrs

(totals must also be entered on IBIS)

For comparison by the examiners only										
Moderator						Senior Moderator				
PE	EX	A	EV	C		PE	EX	A	EV	C
/2	/6	/6	/6	/4		/2	/6	/6	/6	/4
/2	/6	/6	/6	/4		/2	/6	/6	/6	/4

To be completed by the teacher:

Name: **Michael John Dickinson**   Signature: **M. J. Dickinson**   Date: **March 5th 2016**

To be completed by the candidate:

Candidate declaration:
Group 4 Project: The candidate should write a reflective statement of about 50 words outlining their involvement in the group 4 project.

_____
_____
_____
_____

I confirm that the investigation is my own work and is the final version. I have acknowledged each use of the words or ideas of another person, whether written, oral or visual.

Name: **James Bond**   Signature: **J. Bond**   Date: **March 5th 2016**

Sample Form 4/IA
(Internal Assessment Checklist)

> A guess at what the 4/CSS Form might look like has been made for this publication. The actual form will be available in the IB Handbook of Procedures at the start of the 2015-16 school year.

This checklist is produced to help teachers prepare their IA samples for submission to IB for moderation purposes. The checklist itself should not be submitted

---

Please check the boxes to confirm that you have carried out the following requirements in preparing the sample.

- ☐ I have read section XXXX and XXXX and section XXXX in the handbook. (IBO's assessment department will enter the section numbers when known).

- ☐ Internal standardization has taken place where two or more teachers are responsible for the internal assessment of candidates.

- ☐ A form 4/PSOW is included, signed and dated by the teacher.

- ☐ Photocopied material is legible (ideally, original work should be sent to the moderator).

- ☐ The criteria Personal Engagement (PE), Exploration (EX), Analysis (A), Evaluation (EV) and Communication (C) have all been assessed.

- ☐ The levels marked for each of the criteria, PE, EX, A, EV and C have been clearly entered on each candidate's Individual Candidate Cover Sheet (form 4/CSS).

- ☐ The Investigation and accompanying teacher instruction sheets for each candidate in the sample set are enclosed.

- ☐ The title of the Group 4 Project is included in the outline of experiments in the 4/PSOW.

- ☐ Each candidate has written a reflective statement on his/her involvement in the Group 4 Project on the 4/CSS form

- ☐ The experiments/dates on which the students experienced specific ICT applications have been flagged.

- ☐ Each candidate has signed a declaration on academic honesty.

- ☐ The final mark out of 24 for internal assessment has been recorded on each candidate's Individual Candidate Cover Sheet (form 4/CSS) and has been entered on the internal assessment option on IBIS.

Atypical candidates:
It is important that the sample work received by the moderator is typical of the marking standards applied to the whole group of candidates. If IBIS selects a candidate's work for a moderation sample that is atypical, include the work of another candidate with the same or similar mark in addition to the candidate's work.

The Investigation

The Investigation – A New Component of the 4/PSOW

A significant change from the previous syllabus in the way that students' practical skills are assessed is that an individual "Investigation" based on a more protracted, in-depth study of a topic of interest to the student will ultimately constitute the sample sent to the IBO's Internal Assessment Moderator for marking and moderation. This single, 10 hour investigation should be specifically designed to match the relevant assessment criteria described in the following pages. The final report should be about 6 to 12 pages long; investigations exceeding this length will be penalized in the communications criterion as lacking in conciseness. "The Investigation" which constitutes the only internally assessed, externally moderated aspect of the Physics course is worth 20% of the final grade awarded to the student.

Assessment of the Investigation

As stated previously, The Investigation will be the only piece of student work that must be assessed – however, it is advisable to use the marking criteria rubrics formatively for at least some of the other practical activities so that students get a feel for how they are to be assessed before the summative piece is completed and sent to IBO for moderation.

Another suggestion, if time allows, is to consider performing a preliminary Investigation so that students get some practice with a large project before they approach the real one (this might be more plausible with Standard Level groups, depending upon how your school allocates time). Who knows, their preliminary/practice Investigation might be better than the second one.

As part of the learning process, teachers can give general advice to students on a first draft of their work for their Investigation. However, constant drafting and redrafting is not allowed and the next version handed to the teacher after the first draft must be the final one. In assessing student work using the IA criteria, teachers should only mark and annotate the final draft.

[IB Physics Diploma Programme Guide, 2007. ©IBO]

Weighting of the Criteria

The new assessment model uses five criteria to assess the final report of the individual investigation with the following raw marks and weightings assigned:

Criterion	Raw Marks	Weighting
Personal engagement	2	(8%)
Exploration	6	(25%)
Analysis	6	(25%)
Evaluation	6	(25%)
Communication	4	(17%)
Total	**24**	**(100%)**

The mark out of 24 will be converted into a percentage, with a maximum of 20% given to the student. The remaining 80% is allocated to the External Assessment component (examinations). For a more detailed explanation of the Practical Scheme of Work and Internal Assessment requirements please refer to the IB Physics Guide (First Assessment 2016).

Acceptable Levels of Teacher Assistance

At first, it appears that the lab sheets provide very little information to the students. This is by design rather than by accident. As I said in the preface to the first edition of this book, it is important that the lab reports submitted by the students are their own work and not the work of an experienced teacher. Here are some dos and don'ts when it comes to the amount of assistance that you can give to students – especially when it comes time for your students to undertake "The Investigation"

When assessing the Exploration criterion.
- ✓ It is acceptable to give students a general theme to investigate.
- ✓ It is acceptable to give the students the dependent variable (however, doing so might reduce the chances of the student receiving higher marks in this criterion).
- ✗ It is unacceptable to give students both the dependent and independent variables to investigate.
- ✗ It is unacceptable to give students the controlled variables.
- ✓ It is acceptable to show them the range of equipment that your school has for the topic being investigated.
- ✗ It is unacceptable to give students a materials and apparatus list.
- ✓ It is acceptable to explain to students how a piece of apparatus works.
- ✗ It is unacceptable to give students a method / procedure for the collection of data.
- ✓ It is acceptable to teach students how to propagate errors in experiments.
- ✗ It is unacceptable to explain how variables should be controlled (for that specific experiment).

When assessing the Analysis criterion
- ✓ It is acceptable to suggest which variables to collect data for (as long as the experiment is not being assessed for the Design criterion also).
- ✗ It is unacceptable to provide a blank table with headings and units, which is simply to be filled in by the student(s).
- ✓ It is acceptable to suggest that the student(s) process the data in a "suitable" way.
- ✗ It is unacceptable to tell the students to plot a graph of variable A vs. variable B.
- ✗ It is unacceptable to provide pre-labeled graph axes on which students need only to plot points.
- ✓ It is acceptable to suggest that the student(s) investigate the degree of uncertainty in their calculated result
- ✗ It is unacceptable to tell the students exactly how to process the result in order to calculate the uncertainty – they should have already been taught this prior to the investigation.
- ✗ It is unacceptable to provide students with a set of structured questions to be answered.
- ✓ It is acceptable to provide minimal guidance on how a conclusion might be structured.

When assessing the Evaluation criterion
- ✓ It is acceptable to suggest that students compare the result obtained in their experiment to that quoted in the textbook.
- ✗ It is unacceptable to provide students with a set of structured questions to be answered.

INTERNAL ASSESSMENT RUBRICS

Personal Engagement

This criterion assesses the extent to which the student engages with the exploration and makes it their own. Personal engagement may be recognized in different attributes and skills. These could include addressing personal interests or showing evidence of independent thinking, creativity or initiative in the designing, implementation or presentation of the investigation.

Mark	Descriptor
0	The student's report does not reach a standard described by the descriptors below.
1	**The evidence of personal engagement with the exploration is limited with little independent thinking, initiative or creativity.** The justification given for choosing the research question and/or the topic under investigation does not demonstrate **personal significance, interest or curiosity**. There is little evidence of **personal input and initiative** in the designing, implementation or presentation of the investigation.
2	**The evidence of personal engagement with the exploration is clear with significant independent thinking, initiative or creativity.** The justification given for choosing the research question and/or the topic under investigation demonstrates **personal significance, interest or curiosity**. There is evidence of **personal input and initiative** in the designing, implementation or presentation of the investigation.

Exploration

This criterion assesses the extent to which the student establishes the scientific context for the work, states a clear and focused research question and uses concepts and techniques appropriate to the Diploma Programme level. Where appropriate, this criterion also assesses awareness of safety, environmental, and ethical considerations.

Mark	Descriptor
0	The student's report does not reach a standard described by the descriptors below.
1–2	The topic of the investigation is identified and a research question of some relevance is **stated but it is not focused**. The background information provided for the investigation is **superficial** or of limited relevance and does not aid the understanding of the context of the investigation. The methodology of the investigation is only appropriate to address the research question to a very limited extent since it takes into consideration few of the significant factors that may influence the relevance, reliability and sufficiency of the collected data. The report shows evidence of limited awareness of the significant safety, ethical or environmental issues that are **relevant to the methodology of the investigation***.
3–4	The topic of the investigation is identified and a relevant but not fully focused research question is described. The background information provided for the investigation is mainly appropriate and relevant and aids the understanding of the context of the investigation. The methodology of the investigation is mainly appropriate to address the research question but has limitations since it takes into consideration only some of the significant factors that may influence the relevance, reliability and sufficiency of the collected data. The report shows evidence of some awareness of the significant safety, ethical or environmental issues that are **relevant to the methodology of the investigation***.
5–6	The topic of the investigation is identified and a relevant and fully focused research question is clearly described. The background information provided for the investigation is entirely appropriate and relevant and enhances the understanding of the context of the investigation. The methodology of the investigation is highly appropriate to address the research question because it takes into consideration all, or nearly all, of the significant factors that may influence the relevance, reliability and sufficiency of the collected data. The report shows evidence of full awareness of the significant safety, ethical or environmental issues that are **relevant to the methodology of the investigation.***

* This indicator should only be applied when appropriate to the investigation. See exemplars in teacher support material.

Analysis

This criterion assesses the extent to which the student's report provides evidence that the student has selected, recorded, processed and interpreted the data in ways that are relevant to the research question and can support a conclusion.

Mark	Descriptor
0	The student's report does not reach a standard described by the descriptors below.
1–2	The report includes **insufficient relevant** raw data to support a valid conclusion to the research question. Some **basic** data processing is carried out but is either too **inaccurate or too insufficient to lead to a valid** conclusion. The report shows evidence of little consideration of the impact of measurement uncertainty on the analysis. The processed data is incorrectly or insufficiently interpreted so that the conclusion is invalid or very incomplete.
3–4	The report includes relevant but incomplete quantitative and qualitative raw data that could support a simple or partially valid conclusion to the research question. Appropriate and sufficient data processing is carried out that could lead to a broadly valid conclusion but there are significant inaccuracies and inconsistencies in the processing. The report shows evidence of some consideration of the impact of measurement uncertainty on the analysis. The processed data is interpreted so that a broadly valid but incomplete or limited conclusion to the research question can be deduced.
5–6	The report includes sufficient relevant quantitative and qualitative raw data that could support a detailed and valid conclusion to the research question. Appropriate and sufficient data processing is carried out with **the accuracy** required to enable a conclusion to the research question to be drawn that is fully **consistent** with the experimental data. The report shows evidence of full and appropriate consideration of the impact of measurement uncertainty on the analysis. The processed data is correctly interpreted so that a completely valid and detailed conclusion to the research question can be deduced.

Evaluation

This criterion assesses the extent to which the student's report provides evidence of evaluation of the investigation and the results with regard to the research question and the accepted scientific context.

Mark	Descriptor
0	The student's report does not reach a standard described by the descriptors below.
1–2	A conclusion is **outlined** which is not relevant to the research question or is not supported by the data presented. The conclusion makes superficial comparison to the accepted scientific context. Strengths and weaknesses of the investigation, such as limitations of the data and sources of error, are **outlined** but are restricted to an **account** of the **practical** or **procedural issues** faced. The student has **outlined** very few realistic and relevant suggestions for the improvement and extension of the investigation.
3–4	A conclusion is **described** which is relevant to the research question and supported by the data presented. A conclusion is described which makes some relevant comparison to the accepted scientific context. Strengths and weaknesses of the investigation, such as limitations of the data and sources of error, are **described** and provide evidence of some awareness of the **methodological issues*** involved in establishing the conclusion. The student has **described** some realistic and relevant suggestions for the improvement and extension of the investigation.
5–6	A detailed conclusion is **described and justified** which is entirely relevant to the research question and fully supported by the data presented. A conclusion is correctly **described and justified** through relevant comparison to the accepted scientific context. Strengths and weaknesses of the investigation, such as limitations of the data and sources of error, are **discussed** and provide evidence of a clear understanding of the **methodological issues*** involved in establishing the conclusion. The student has **discussed** realistic and relevant suggestions for the improvement and extension of the investigation.

*See exemplars in teacher support material for clarification.

Communication

This criterion assesses whether the investigation is presented and reported in a way that supports effective communication of the focus, process and outcomes.

Mark	Descriptor
0	The student's report does not reach a standard described by the descriptors below.
1–2	**The presentation of the investigation is unclear, making it difficult to understand the focus, process and outcomes.**
	The report is not well structured and is unclear: the necessary information on focus, process and outcomes is missing or is presented in an incoherent or disorganized way.
	The understanding of the focus, process and outcomes of the investigation is obscured by the presence of inappropriate or irrelevant information.
	There are many errors in the use of subject specific terminology and conventions*.
3–4	**The presentation of the investigation is clear. Any errors do not hamper understanding of the focus, process and outcomes.**
	The report is well structured and clear: the necessary information on focus, process and outcomes is present and presented in a coherent way.
	The report is relevant and concise thereby facilitating a ready understanding of the focus, process and outcomes of the investigation.
	The use of subject-specific terminology and conventions is appropriate and correct. Any errors do not hamper understanding.

*For example, incorrect/missing labeling of graphs, tables, images; use of units, decimal places.

[IB Physics Diploma Programme Guide, 2014. ©IBO]

Sample Investigation
Hot-Air: Investigating the uplift of a hot air balloon

Silvio Augustin, UNIS Hanoi, 2006

Part 1 – Exploration

Introduction

I have always been interested in, and curious about Hot-Air Balloons. Throughout the history of mankind, the great dream was to fly. Attempts of jumping off church tops in a bird outfit did not always end successfully. To achieve enough uplift to stay in the air by imitating birds is one method of flying that requires a lot of energy and a large surface area to displace air, but in fact, there is a much simpler way to see the clouds from above. The first pioneers of ballooning were the Montgolfier brothers. After a series of unsuccessful experiments to build and perform a balloon take off they created an 800 m³ balloon which rose to a height of 1,000 meters, and traveled a distance of two kilometers. Although the brothers succeeded building a hot-air balloon they did not understand the physics of it very well. The smoke that was made by the fire was considered to be the cause of the uplift. The denser the smoke, the better – was their theory. A fire from straw, humidified wool and even of old shoes made the flight a success! Later the French physicist Jacques Charles discovered that the volume of an ideal gas is proportional to its temperature, which means that hot air has less mass per unit of volume than cold air.

$$\frac{T}{V} = k$$

This phenomenon allows a hot air balloon to rise in colder air and enable humans to travel over long distances. Since I was always very interested in flying transportation methods, and especially those which seem so fundamental as hot air balloons, I found this essay a great opportunity to extend my knowledge of hot air balloons and of the relationship between it variables.

My theoretical and experimental investigation is based on the question:

What is the effect of a temperature differential between the walls of a hot air balloon on the uplift force?

Figure 1: Piccard and Brian Jones Betrand, first men to circumnavigate the globe in a hot air balloon.

Considering Charles's Law, I would hypothesize a cubed relationship between the named variables. I believe that when the volume doubled as the temperature (in Kelvin) doubled the density must have diminished by a factor of 3. Density is the same as the mass per the unit of volume. therefore the cubed uplift force is proportional to volume.

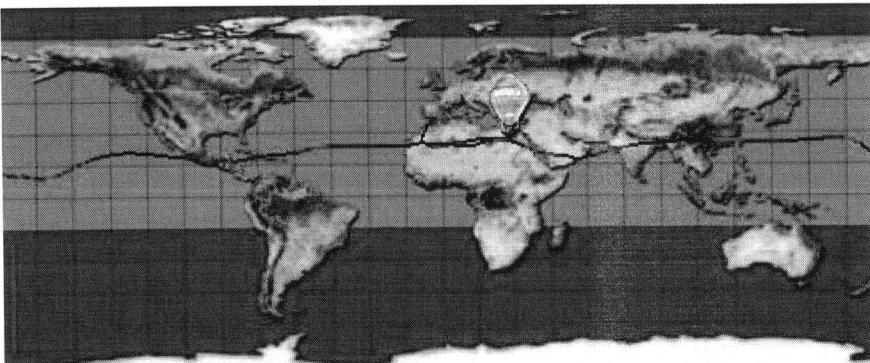

Figure 2: Around the world in 19 days

This simple physical law enables humans to fly very long distances and not so long ago circumnavigate the earth. The two men, Piccard and Brian Jones Betrand, started their journey from Switzerland and drifted along the equator for 19 days until they arrived at their destination, which was their origin at the same time.

Figure 3: Breitling Orbiter 3 was powered in the vertical simply by the temperature difference and along the equator by the convection currents that exists due to the same principal - hot air rises in cool air.

Comparison of Model and Passenger balloon
1. My Model Balloon

Envelope
In our case, the envelope is made out of a very thin PE-foil. Its maximum temperature tolerance is around 200°C. The material is very light, the 3 g foil manages to hold about 22 dm3 of air.

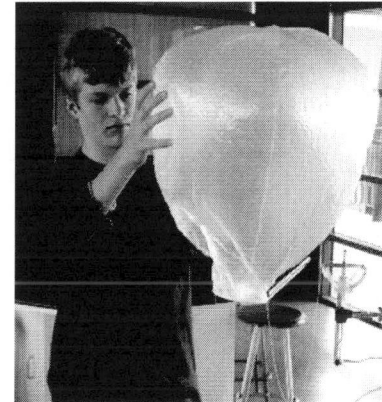
Figure 4: Model balloon

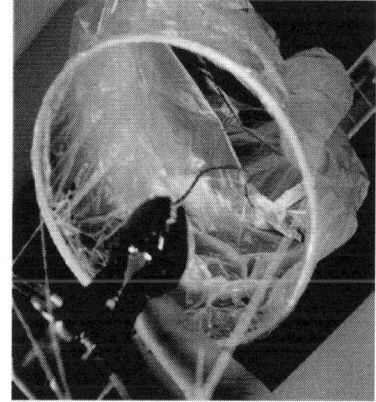
Figure 5: Fuel cell sitting on a clamp, heating the inside of the balloon

Skirt
The ring of thin wire, which can be associated with the skirt, keeps the balloon mouth open so the heated air from the fuel cell can enter the balloon without melting the side.

Burner
The fuel cell is wrapped into an aluminum foil and is open at the top. It burns for about 60 seconds at a temperature of 800°C while it is hanging on a hook extended from the skirt.

Figure 6: the fuel cell – before and after

2. Passenger Balloons

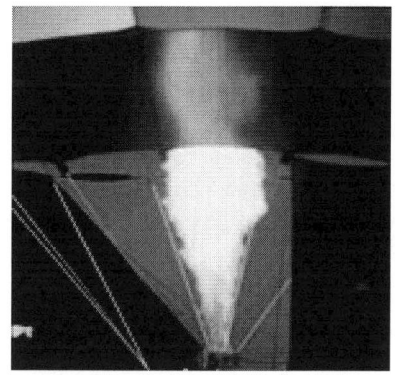

Envelope
The envelope is the actual balloon part. It is made up of long nylon sections called gores. As we see in figure 7, the gores are sewed together and reach from the bottom to the crown of the balloon. The nylon polyester combination has a very low weight and is an appropriate material due to its high melting point. It is very tightly woven and coated with a material which makes it air tight and durable.

Figure 7: Passenger balloon   **Figure 8:** Burner and skirt

Skirt
The skirt is referred as the bottom of the envelope; it is made out of the material known as "Nomex". "Nomex" is flame resistant and is also used in Formula 1 and Firefighting equipment. "Nomex" is vital to a functional balloon since it is used near the mouth of the envelope where the propane burner flame is directly exposed to it.

Burner
The balloon of today uses propane fuel to heat the air in the inside of envelope. The fuel is kept compressed in cylinders which are stored in the basket. The pilot can control the strength of the propane flame. When one wants to increase the floating height, one has to turn up the flame. The rings around the burner heat up as the flame burns. As the propane liquid flows through the burner ring-tubes it heats up and turns into a gas which increases the overall fuel efficiency since the gas provides a more powerful flame compared to the liquid.

Figure 9: Burner

Figure 10: The basket

Basket
The basket is hanging onto the skirt of the balloon. Some carry up to 20 passengers or even more, plus pilot. On top of that, they carry the propane cylinders.

My Research Question:

What is the effect of a temperature differential between the walls of a hot air balloon on the uplift force?

Background Research
Air Pressure + Gravity = Buoyancy.
How does temperature and gravity affect air pressure and what is the relationship between those and the uplift force of a hot air balloon? To answer this question we need to step back and consider the forces acting around us. Every mass is attracted to the center of the earth by the gravitational force. Therefore also every gas molecule is pulled downward. However, we have never experienced that air molecules gather together on the bottom of a room. But if we would consider the earth surface as the bottom of a room, we see what effect gravity has on the mass of a large gas volume. At sea level air pressure is relatively high but as we travel in altitude the atmospheric pressure decreases.

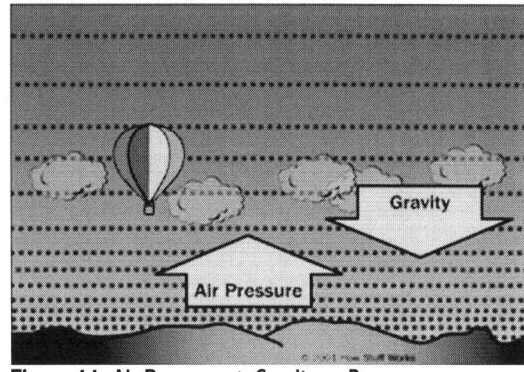

Figure 11: Air Pressure + Gravity = Buoyancy

We know that any gas molecule moves rapidly in random movement. Many molecules fill out the container they are in. A molecule changes direction when it collides with another molecule or hits the wall of its container. Although there is no wall stopping the gas molecules to escape into outer space, nevertheless we are still able to move in an ocean of air molecules.

Therefore, we can conclude that "All air particles in the atmosphere are drawn by the downward force of gravity. But the pressure in the air creates an upward force working opposite gravity's pull. Air density builds to whatever level balances the force of gravity, because at this point gravity isn't strong enough to pull down a greater number of particles."

[From: http://travel.howstuffworks.com/hot-air-balloon5.htm]

The reason that molecules of a substance in gaseous state travel so rapidly is their high kinetic energy. When the temperature increases, the molecules get more "excited", move faster and expand in all directions or build up pressure in the container they are in. Therefore temperature is an indicator of the average kinetic energy for any volume of substance in gas form. When the gas molecules gain in kinetic energy they collide more frequently, building up pressure. Considering a fixed mass, and pressure staying constant, the gas will expand and gain in volume as the temperature rises. (See next graph).

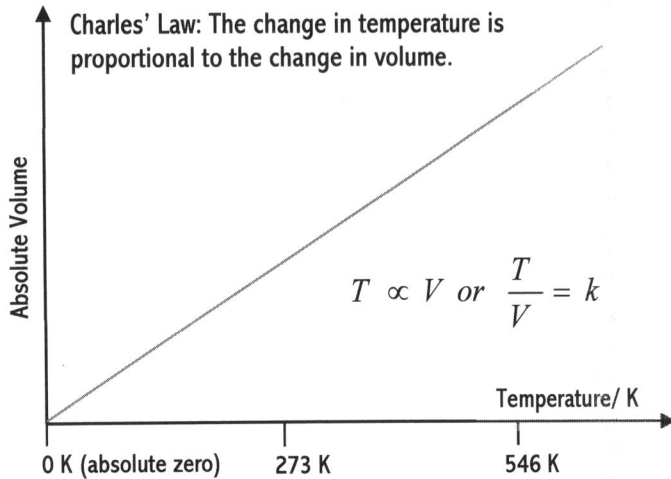

Charles' Law: The change in temperature is proportional to the change in volume.

$$T \propto V \text{ or } \frac{T}{V} = k$$

Why does a hot air balloon fly?
We know so far that gravity and the molecular kinetic energy are in equilibrium. And we can state that hot air requires more volume than cold air and therefore has less mass per unit of volume which makes it lighter than cool air.

The result is that the heavier cool air sinks due to gravity and pushes the lighter warm air up at the same time. The hot air that is trapped inside of the balloon can only expand downwards which leaves less dense air behind that is moving upward. As a result the hot air balloon rises.

As we can see on the data table below the mass of one m^3 of air is 1.293kg at 0°C. As we heat up the air to a temperature of 100°C the air only has a mass of 0.946 kg which is 347g less than at 0°C. For example, a balloon with a volume of 20m^3 can lift a weight of about 6.9 kg if the temperature difference is 100 °C! This might not seem that much but that is why hot air balloons have such a large volume.

Temperature / t (°C)	Density / ρ (kg/m³)
-40	1.514
-20	1.395
0	1.293
20	1.204
40	1.127
60	1.060
80	0.9996
100	0.9461
200	0.7461
500	0.4565
1000	0.2772

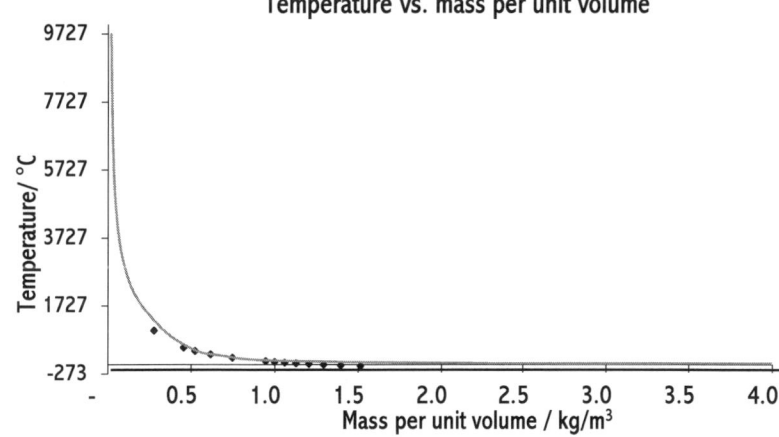

From: [http://www.engineeringtoolbox.com/air-desity-specific-weight-22_600.html]

As we can see from the following graph, the relationship is asymptotic. As the temperature approaches absolute zero the mass per unit of volume approaches infinity and as temperature reaches infinity the mass per volume is approaching zero. (Absolute zero = –273 °C)

Conclusion:
From my investigation I understand that the existence of gravity is vital for the uplift of a hot air balloon. The gravitational field builds up an ocean of gas around the earth. Gas particles have the physical properties to fly around freely in random movement at a high velocity. Their high kinetic energy forces the molecules to spread around and fill out their container which is the universe. However gravity and the air pressure are in equilibrium therefore no more air molecules can escape the gravitational field of the earth. As we increase the temperature of a container of gas the average kinetic energy will increase forcing the molecules to move faster, to collide more frequently and to expand proportionally in volume. As a result the mass per unit of volume would be spread out in all 3 dimensions giving us a cubed relationship to temperature.

Part 2 - My Experiment

Introduction:
After investigating the forces acting on a hot air balloon, I decided to proceed with practical work. By letting a hot air balloon model hover, I wanted to observe the relationship between temperature difference and the uplift force. I devised a plan how to measure the small forces exerted on the balloon. My original idea was to build a couple of Hot-air balloons myself but I reconsidered and came to the conclusion that my experiments would have more consistency if my hot-air balloons had consistent weight and form. Accordingly I bought 4 sets of hot air balloons including fuel cells during a winter stay in Berlin via Internet. The balloons are made out of thin plastic foil that has a maximum temperature resistance of about 250°C. The fuel cells (4 per balloon) had a life span of about 60 seconds and reached 900°C at the hottest point of the flame. Accordingly the flame and the plastic foil had to stay in a safe distance of about 2-3 cm. The experiments cost me 2 of my 4 balloons. The following pages show my failures and successes.

Experiment 1:
The original idea was to set up the experiment as shown to the right. The force meter, attached to the floor and to the clamp stand, would be giving the upward force and the 2 thermometers reading the temperature difference. However even before I started I realized that the small balloon and the low temperature difference would not generate enough uplift to fulfill the force meter's purpose. The scale was a 10th of a Newton per division and therefore too large for our purposes.

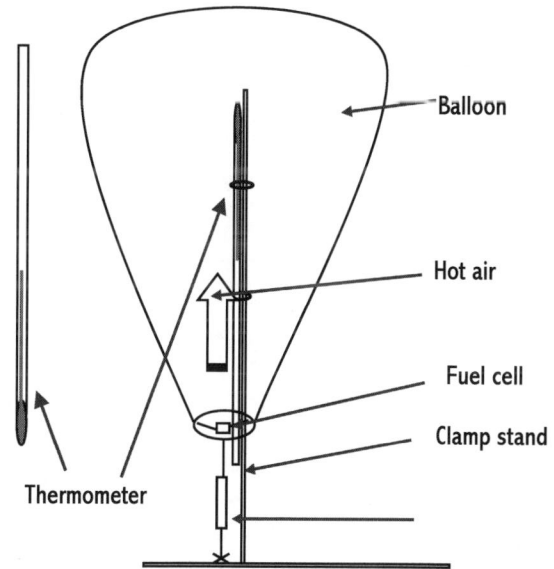

Figure 12: Experiment 1 Setup

Experiment 2:

After considering the dilemma I came up with an alternative indirect force measure. I tied a very light cotton string to the hot air balloon and connected the other end with a 10g weight that I placed on a high sensitive balance where I could read of the decreasing weight. Now this method would give me valuable force readings but the temperature would have been unknown. The temperature measurement was probably the most difficult obstacle in this experiment. In my original plan I introduced the idea that I clamped a thermometer upside down onto a clamp stand but in fact the balloon's opening is too small to fit a clamp stand plus a thermometer clamped to it.

I overcame this problem by simply taping the thermometer to the clamp stand instead. But yet again we had a problem- friction. As one can see in figure 15 the digital balance has to stand next to the clamp stand and the strings pull the balloon towards the balance while the ring around the balloon's mouth rubs against the stand. This causes relatively high amounts of friction considering the very sensitive measurements that are in progress.

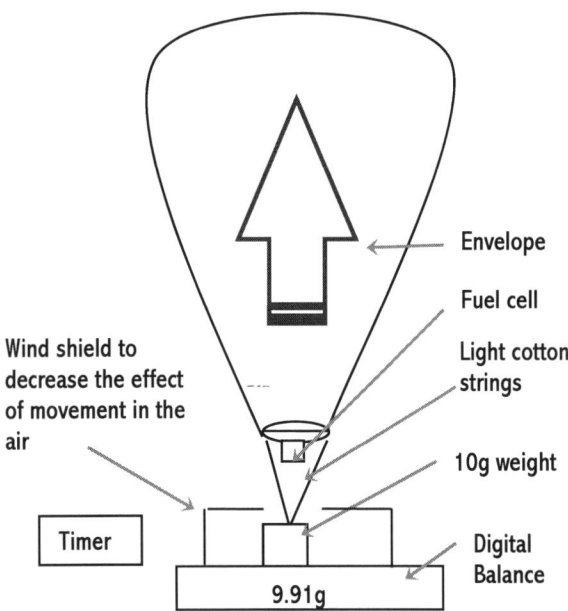

Figure 13: Experiment 2 Setup

A possible solution for this problem is to perform the whole experiment in steps. First measuring the upward force with the digital balance from the point on when the balloon has enough inside temperature to hover in the air till it starts falling again. The time needs to be taken parallel to the force measurement starting at the moment the balloon hovers till it drops again.

Then start the experiment again and measure the temperature during the same time periods. The results of the two steps could be drawn together on one graph. However this way of performing the experiment would not give us accurate results due to changing conditions. If I would use the same balloon the balloon would have a weight difference after the first step.

The exhaust from the fuel cell and the water vapor that is hanging in the envelope forbids the balloon to hover again. If a different balloon would be used the volume and weight would also differ slightly.

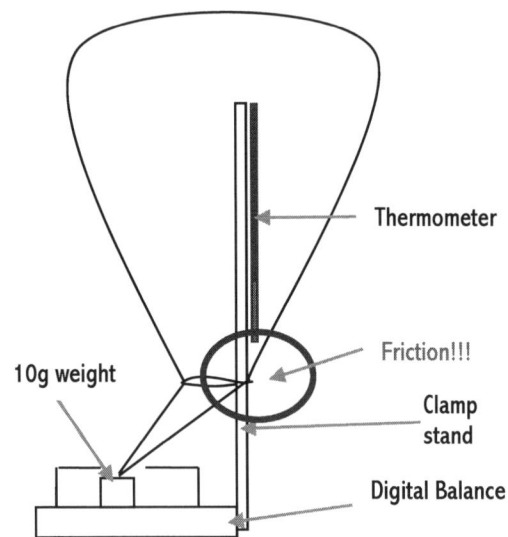

Figure 14: Experiment 2 Setup - Modified

Experiment 3:

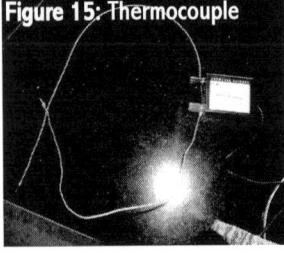

Figure 15: Thermocouple

Figure 16: Digital force meter

Thankfully, life is full of surprises sometimes! My Teacher introduced two apparatus to me that are linked to a lap top which give me data at a high frequency and accuracy. Instantly I gave up my plan, performing the experiment in steps and designed a new plan including the new technology, a thermocouple shown in figure 15. In order to measure the small upward force, the highly sensitive force meter is required. In Figure 16, the force meter is shown. Two strings are attached to the balloon's skirt, which lead down to the hook (figure 17). The force pulling on the hook is recorded on the computer. The thermocouple is also shown on the picture. It can be attached to the computer and gives us a reading of the temperature difference between the two ends of the strings. Figure 17 also shows that one string leads into the balloon and one string remains outside. As the experiment is taking place, the computer gives the temperature difference between the inside of the balloon and the outside.

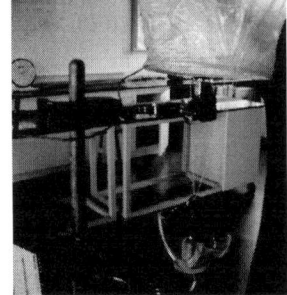

Figure 17: New Setup

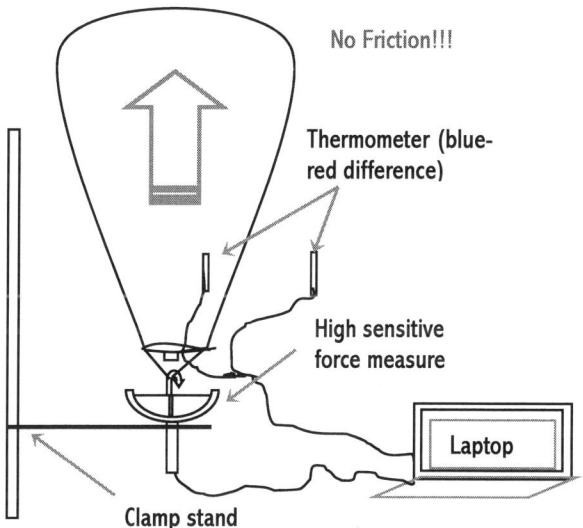
Figure 18a:

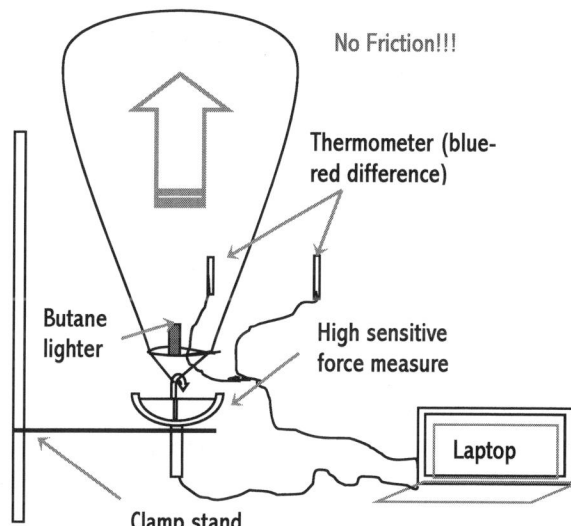
Figure 18b:

Data Collection:
My final experiments at last gave me some reasonable results. However I had to manipulate quite a lot which made my information rather random. By looking at my graphs I tried to remove possible outliers. The results table shows the data without my identified outliers.

Time, t [s]	Temperature, T [°C] ±0.0001	Uplift [N] ±0.00001	Temperature Uncertainty %	Uplift Uncertainty %
1	14,861	0,1836	0,00067%	0,0054%
2	15,556	0,1975	0,00064%	0,0051%
3	16,257	0,2138	0,00062%	0,0047%
6	17,639	0,2069	0,00057%	0,0048%
8	18,333	0,2069	0,00055%	0,0048%
9	19,028	0,2092	0,00053%	0,0048%
10	23,194	0,2092	0,00043%	0,0048%
11	20,417	0,2092	0,00049%	0,0048%
12	20,417	0,2115	0,00049%	0,0047%
14	21,806	0,2255	0,00046%	0,0045%
15	27,361	0,2138	0,00037%	0,0047%
20	26,667	0,2302	0,00037%	0,0043%
22	28,056	0,2255	0,00036%	0,0045%
24	30,139	0,2255	0,00033%	0,0044%
26	32,222	0,2302	0,00031%	0,0043%
27	33,611	0,2302	0,00030%	0,0043%
28	36,389	0,2302	0,00027%	0,0043%
29	33,611	0,2115	0,00030%	0,0047%
31	37,778	0,2534	0,00026%	0,0039%
32	34,306	0,2348	0,00029%	0,0043%
39	42,639	0,2534	0,00023%	0,0039%
41	44,722	0,2441	0,00022%	0,0041%
42	42,639	0,2674	0,00023%	0,0037%
43	46,806	0,2488	0,00021%	0,0040%
45	50,278	0,2558	0,00020%	0,0039%

In my experiments I used digital measuring devices that had a high accuracy. As we can see on the data table there are many numbers from both of the data sets that are exactly the same. I suspect that the digital measuring devices are actually not that sensitive as the decimal places suggest. A force measuring device that measures accurate data down to 4 decimal places should obtain maybe similar data but not exactly the same value 4 successive times. I have highlighted numbers on the data table that are the same value. The repeating numbers suggest to me that the force measuring device has an uncertainty of ±0.001 but not ±0.00001. As well the thermocouple has repetition in its values, I suggest an uncertainty of ±0.001 instead of ±0.0001. However those uncertainties are not significant enough to include them on any graphs.

Part 2 - Analysis

The whole data table gives me very scattered points that do not show me a clear relationship between the temperature difference and uplift force. I looked at my results for the temperature time graph and the results for the uplift time graph and discovered that there is an obvious trend line, however quite a few outliers. Both graphs are plotted below. I circled the outliers that I took out of the graph.

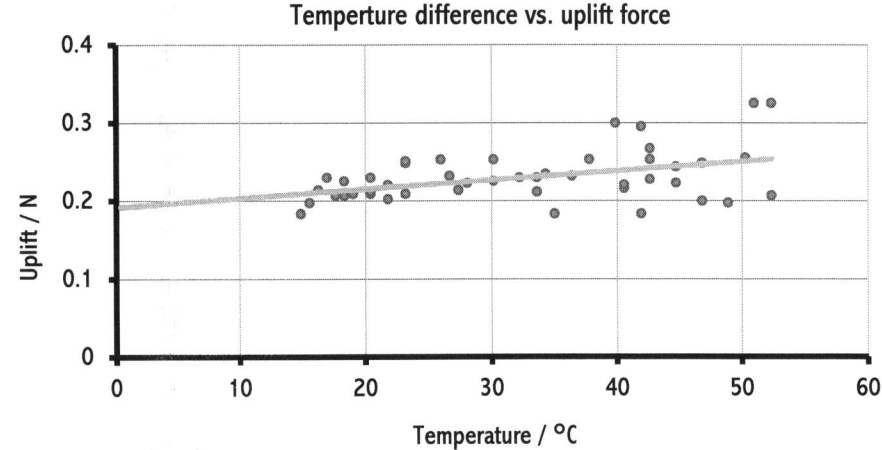

I circled and removed the outliers on the Uplift vs. time graph taken in the following seconds: 5, 7, 13, 16, 18, 19, 23, 33, 35, 36, 37, 38, 44, 46, 47. I circled the data on the temperature/time graph taken in the following seconds: 4, 16, 17, 18, 21, 23, 25, 30, 34, 40.

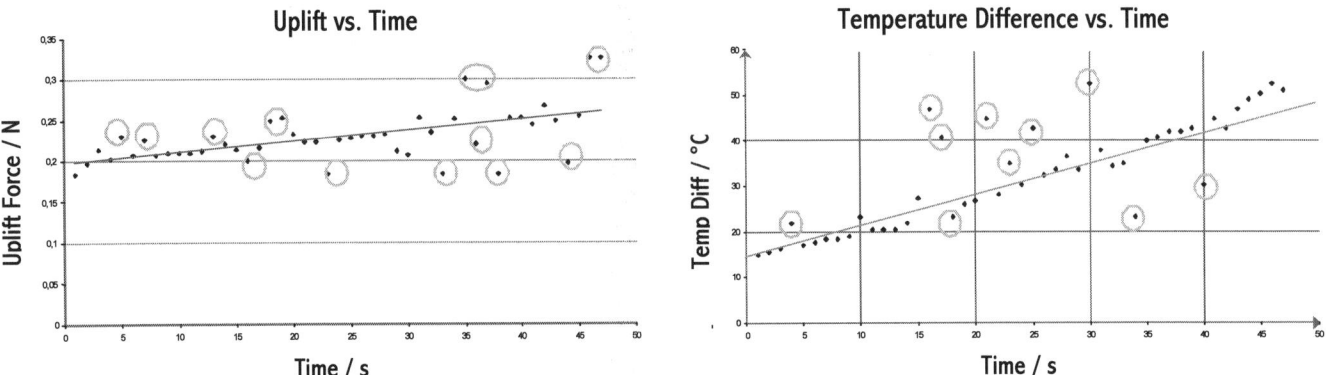

After removing the outliers from my original data table I plotted the next graph that shows the Temperature vs. Uplift graph after I took out both the outliers from the uplift and the temperature graph for all the seconds critical in either of them. The adjusted data table is shown at the top of this section.

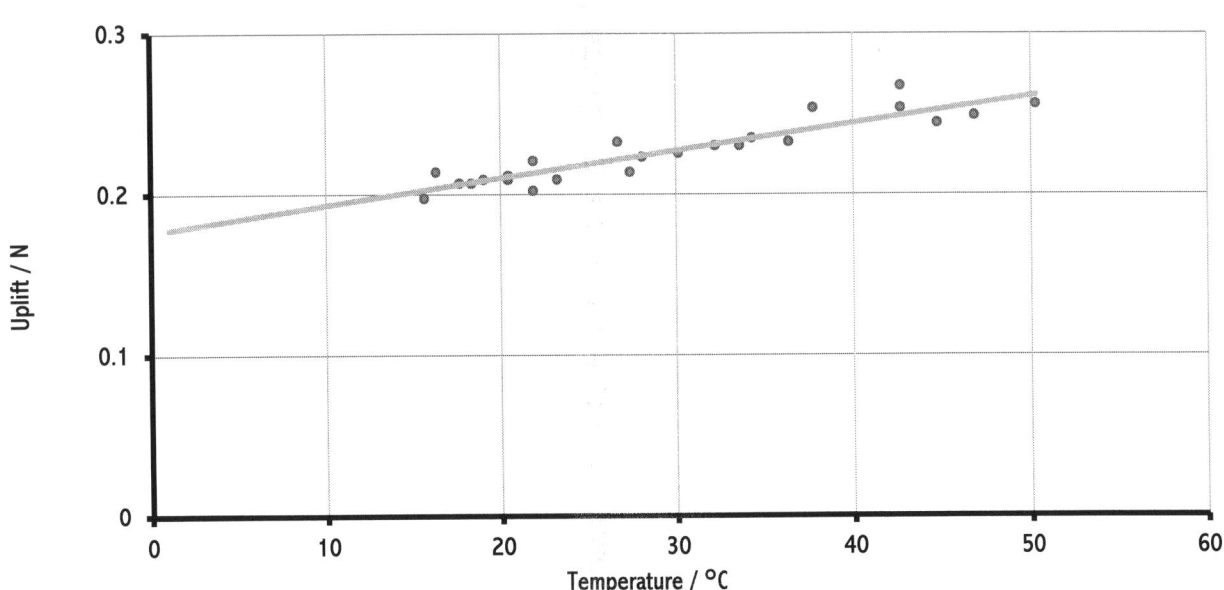

After we have derived my original data to a reasonable graph which has a linear relationship we can find the gradient of the line.

$$Gradient = \frac{rise}{run}$$

$$Gradient = \frac{y_2 - y_1}{x_2 - x_1}$$

$$Gradient = \frac{0.270 - 0.175}{50 - 0} = \frac{0.095}{50} = 0.0019 \left[\frac{N}{\Delta °C}\right]$$

To that value we need to add the weight of the balloon since the measuring only started when the balloon lifted off, hence the uplift force equals the weight of the balloon plus the data from the force meter. The balloon had a weight of 4.94 g giving us a specific weight of 0.0485 N (0.00494 kg x 9.81N)

Therefore our new gradient is 0.0504 N / °C

I know that the gradient value that I obtained is the uplift per one degree temperature difference for the volume of my model balloon. We could idealize the balloon as an ellipsoid and calculate its volume: The balloon has a vertical diameter of 40 cm and a horizontal diameter of 32 cm.

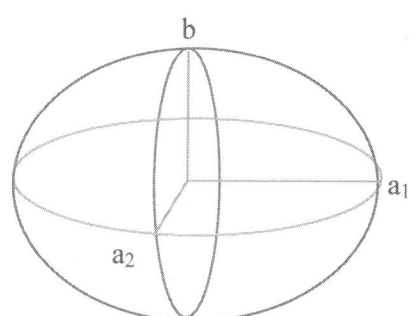

Figure 19: Ellipsoid Balloon

$$V = \frac{4}{3} * \pi * b * a_1 * a_2$$

$$V = \frac{4}{3} * \pi * 16 * 20 * 16$$

$$V = 21390.00 cm^3$$

$$V = 21.39 dm^3$$

Therefore the volume of my model balloon is about 21.39 dm³ which is the same as 0.02139 m³. Now we can convert my gradient from the current form which is $\left[N\Delta°C^{-1} 0.02139 m^{-3}\right]$ to $\left[N\Delta°C^{-1} m^{-3}\right]$.

$$\frac{1}{0.02139} = 46.75$$

$$gradient \times 46.75 = 0.0504 \times 46.75 = 2.354 \left[N\Delta°C^{-1}m^{-3}\right]$$

Now we have calculated the constant for this experiment. The upward force increases by 2.354 N per 1 °C temperature difference for a volume of 1 m³.

Argument/ Data Evaluation
The trend line shows us a linear relationship different from what I have expected. There are reasonable arguments that the relationship between Uplift and temperature is of a cubed manner. That means as the temperature increases the Uplift force increases by a factor of 3. As I already explained in section 4.1.1, volume is proportional to temperature. As the mass of a unit of volume has to be spread out in 3 dimensions the density decreases. Because volume is proportional to temperature, mass³ is inversely proportional to volume. Therefore, the mass³ is also inversely proportional to temperature. The uplift force³ being the mass difference between the volume of air is therefore proportional to the temperature.

$$\because V \alpha T$$
$$\therefore V \alpha M^3$$
$$\therefore T \alpha M^3$$
$$\therefore \Delta T \alpha (Nkg)^3 = (F_{uplift})^3$$

The literature values that I have used earlier in this report agree with this phenomenon.

Δ Temperature Initial T =0°C [°C]	Δ Density [kgm^{-3}]	Uplift Force [N]
0	0,000	0,000
20	0,089	0,089
40	0,166	0,166
60	0,233	0,233
80	0,293	0,293
100	0,347	0,347
200	0,547	0,547
300	0,677	0,677
400	0,769	0,769
500	0,837	0,837
1000	1,016	1,016

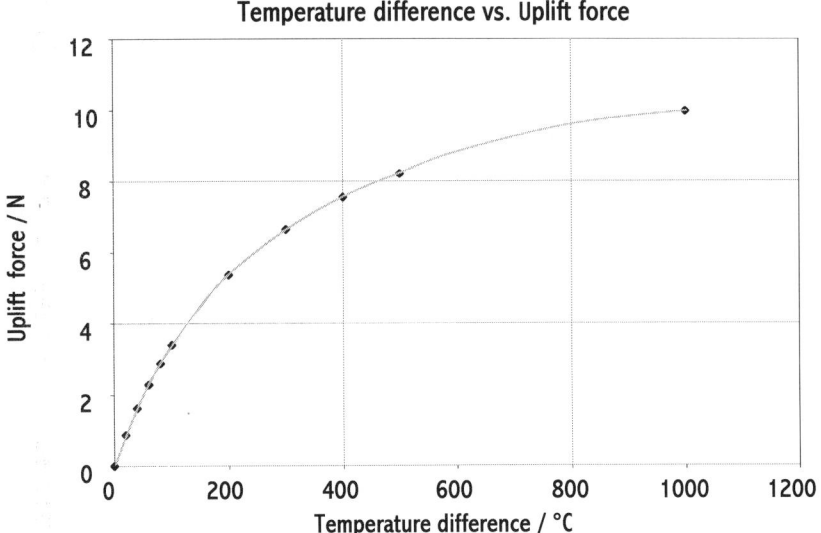

My graph is clearly showing a linear relationship however, we could interpret the graph differently. Before the dotted line the balloon did not have enough uplift to hover in the air. Therefore we don't have any data in that area. The circled data points might have seemed like outliers on my linear trend line, but with a cubed trend line they seem to hit right on the line. Therefore a cubed trend line as shown on the graph below is not out of the question.

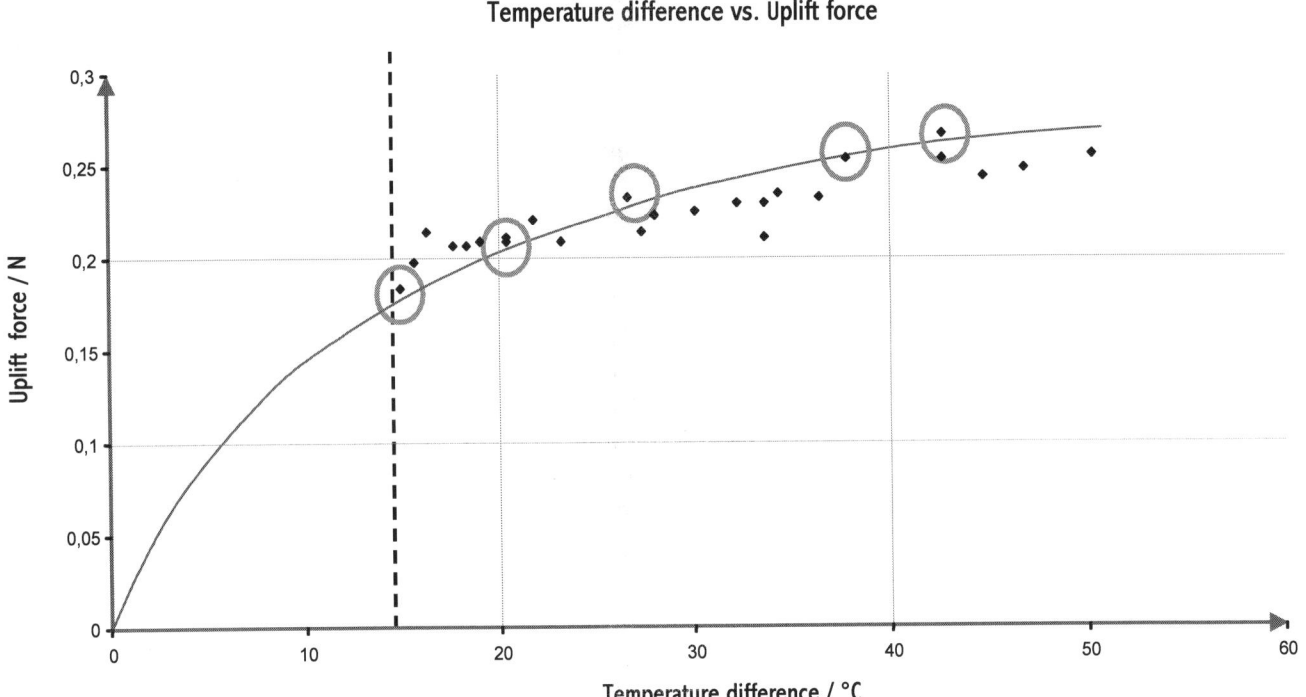

Conclusion:
I have come to the end of my investigations. I explained the principals that enable hot air balloons to fly. I have performed experiments with a model hot air balloon that were an attempt to prove the theory again. The data that I have recorded did not show us the expected relationship however there is some evidence that the graph could have been of a cubed nature but not having a wide enough data range made it difficult to prove this. The gradient obtained by the supposed straight line can be considered useless because the large mass of the model balloon, the small scale of the experiment and the small forces made it difficult to obtain the full picture that was hoped for. I have suggested a new plan that might achieve this wide data range, but for the moment, we need to be satisfied with the fact that hot air moves up and cold air drops down.

Evaluation and Further Analysis:

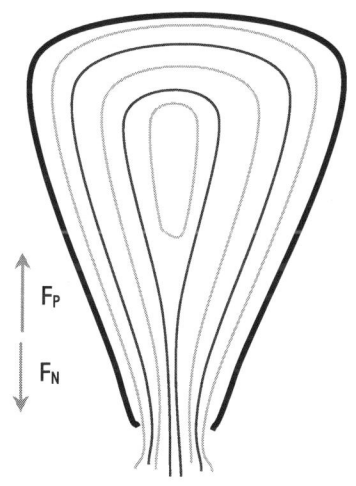

Figure 20: Equilibrium

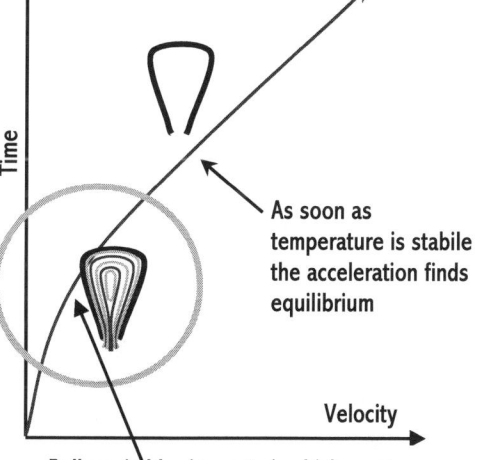

I believe that the upward acceleration for a hot air balloon is not constant throughout the flight. There are stages where the acceleration of the balloon is higher than in others temperature staying constant. Let us consider the following scenario: We have a hot air balloon where in and-outside temperature is equal. We start with heating the inside. We know that the pressure inside the balloon is at constant and we also know that the volume of air is proportional to the temperature therefore the air inside the balloon is expanding. So the air does not fit inside the balloon anymore and it streams out of the balloon's mouth to equalize the pressure between the walls of the balloon! So we can state that there are actually 2 forces acting upward onto the envelope. This downstream of air is like a upward push additionally to the density fall and happening every time the temperature rises.

From the equation of continuity we understand that the air streaming out of the balloon will gain in velocity as it has to pass by the small mouth of the balloon. From that we could understand how much force is exerted upwards.

Further Improvements:

Reconsidering my research question, I realised that I am basically finding the relationship between density (mass per unit of volume) and temperature. Bearing that in mind I could transfer from the balloon to a same volume metal container.

Instead of the 21.39 litre balloon I could use a 21.39 litre ball or box made out of an inflexible fire prove substance. A metal would be the easiest. A pressure valve would allow air to flow out but not in. The air density inside the container decreases as the temperature increases. The temperature could be controlled by changing the height of the flame or the distance to the flame. The force measure sends the decreasing force values to the lap top. The temperature could be taken with a portable thermometer.

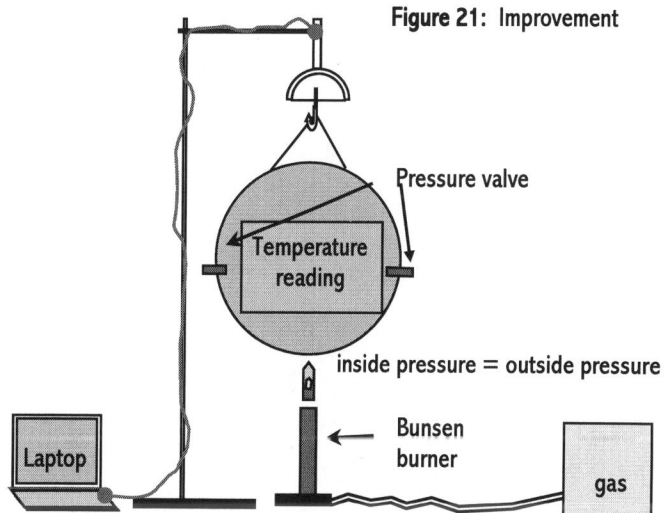

Figure 21: Improvement

With this experiment I could measure the effect on density from the very first moment the temperature rises. I could reach much higher temperatures giving us a broader range of data to find the relationship between temperature and uplift force.

Bibliography:
Giancoli, Douglas C. Physics Principals with Applications. Prentice Hall, Upper Saddle River, New Jersey. 1998
The Engineering Toolbox. Air - Density and Specific Weight. <http://www.engineeringtoolbox.com/air-desity-specific-weight-22_600.html>
Harris, Tom. How Hot Air Balloons Work. 18 March 1998, <http://travel.howstuffworks.com/hot-air-balloon.htm>.

Acknowledgements:
A big thank you to my Teacher, Mr Dickinson for reassuring advice and a steady hand during my experiments.

Physics Practical Scheme of Work – For use with the IB Diploma Programme – First Assessment 2016

MARKING THE SAMPLE INVESTIGATION

Personal Engagement:
- Evidence of personal engagement:
- Justification given for choosing topic:
- Personal significance, interest or curiosity:
- Evidence of personal input and initiative:

The student expressed (minimally) an interest and curiosity about the physics of hot-air balloons early in the essay, however, the justification for choosing the topic is missing. There are numerous instances of significant personal engagement and it appears obvious that the student was enjoying the experience that the investigation offered; initiative and creativity were present when challenges in the data collection stage of the investigation presented themselves. On the whole, there is enough to warrant the awarding of a level 1, but certainly not enough for the student to earn full marks for the Personal Engagement criterion. Mark awarded 1/2.

Exploration:
- The topic of the investigation is identified:
- Stated, described and focused research question:
- Relevant background information provided:
- Background information enhances the understanding of the context of the investigation:
- Appropriate methodology of the investigation:
- Factors affecting relevance, reliability and sufficiency of the collected data:
- Awareness of the significant safety considerations:

The topic of the investigation is clearly identified and the research question is clearly stated. Background information is provided regarding the historical developments in hot-air ballooning, but this background information does not extend to the relevance of the research question; nor does the background information fully enhance the reader's understanding of the context of the investigation. Methodologies employed by the student in investigating the research question are sound and there is a clear development of the experiment after setbacks and unforeseen difficulties were encountered. There is minimal evidence of the student's appreciation of the factors affecting relevance, reliability and sufficiency of the collected data, and more detail could have been added in this area (for example, the removal of outliers in the data were done by eye rather than by using a mathematical method, such as uncertainty bars). Safety consideration were not mentioned by the student and, in the photograph of him holding the hot-air balloon, he appears not to be wearing eye protection. The achievement level is somewhere between 3 and 4; the student has been given the benefit of doubt and the best-fit model of assessment has been used. Mark awarded 4/6.

Analysis:
- Sufficient relevant quantitative and qualitative raw data:
- Data supports detailed and valid conclusion:
- Appropriate and sufficient data processing is carried out:
- Accuracy of data processing allows a conclusion to be drawn consistent with the experimental data:
- Appropriate consideration of measurement uncertainty:
- Correct interpretation of the processed data:
- Valid and detailed conclusion drawn, consistent with the experimental data:

The student recorded sufficient quantitative raw data that allowed a detailed and valid conclusion to be drawn. The student processed the data appropriately and accurately which lead to a conclusion to be drawn consistent with the experimental data. Measurement uncertainty was mentioned but not fully addressed. An attempt was made to interpret the processed data, but the lack of data points at the lower end of the final graph meant that some guesswork was made on the graph's correspondence to the suggested (cubed?) relationship, but the student does allude to this uncertainty in his conclusion. A reasonable attempt has been made to interpret the data such that a valid and detailed conclusion to the research question could be suggested. The achievement level is somewhere between 4 and 5; the student has been given the benefit of doubt and the best-fit model of assessment has been used. Mark awarded 5/6.

Evaluation:
- Conclusion is described and justified:
- Conclusion is relevant to the research question:
- Conclusion is supported by the data presented:
- Conclusion is justified through relevant comparison to the accepted scientific context:
- Strengths and weaknesses of the investigation are discussed:
- Limitations of the data and sources of error, are discussed:
- Evidence of an understanding of methodological issues:
- Realistic and relevant extension of the investigation is suggested:

The conclusion, while both valid and relevant to the research question, is not as detailed as it could be. The student has attempted to make a conclusion that is supported by the data, but acknowledged that the data is somewhat inconclusive. The student attempted to justify the conclusion through relevant comparison to the accepted scientific context, but understands that this is somewhat flawed. Strengths and weaknesses of the investigation are mentioned, but not discussed to any real depth. Limitations of the data and sources of error are glossed over a little. The student demonstrates an understanding of the methodologies used and those that might have been used in the "further improvement" section for his report. This section also suggests a realistic and relevant extension of the investigation. Mark awarded 4/6.

Communication:
- Presentation of the investigation is clear:
- The report is well structured and clear:
- Coherent presentation of focus, process and outcomes:
- The report is relevant and concise:
- The report facilitates an understanding of the focus, process and outcomes of the investigation:
- Appropriate use of subject-specific terminology and conventions:

The final report is well structured and presented in a clear and coherent way which facilitates an understanding of the focus, process and outcomes of the investigation. There is consistency in titles, headings, labeling of figures and page numbering. The report is divided into appropriate sections which aid the reader's appreciation of the scope and sequence of the investigation. While the use of English is not perfect, there is an appropriate use of subject-specific terminology and conventions, with a few errors – these errors do not detract from the reader's understanding of the report. A bibliography has been included. Mark awarded 3/4.

Total Mark Awarded: 17/24

Required Experiments

Prescribed Experiments / Practical Experiences

The new Physics syllabus contains a number of prescribed experiments that must be completed by all students. These experiments are not explicitly listed in the subject guide, but instead they appear in the "Application and Skills" section of each topic's outline (unit of study) in the left hand column of the topic's guide. These experiments and practical activities are generic rather than specific, so the details of implementation are up to the teacher. These prescribed investigations correspond to those experiments often regarded as good practice found in the practical schemes of work over recent years from IB schools.

These generic experiments are required and should appear on the school's 4/PSOW. Questions relating to these may appear on the examination and so it is vital that all students have been exposed to these in order to be fully prepared for the exams. The experimental investigations that appear on the right-hand side of the syllabus format under Aim 6 are not required but serve to help the teacher in producing a strong 4/PSOW. Here is a summary of the left-hand side application and skills prescribed experiments.

List of Prescribed Experiments / Practical Activities

Topic 2.1 - Determining the Acceleration of Free-Fall.

Topic 3.1 - Applying the calorimetric techniques of specific heat capacity or specific latent heat.
(Choose at least one from the following titles)
- Investigating Specific Heat Capacity by the Electrical Method.
- Investigating Specific Heat Capacity by the Method of Mixtures.
- Investigating Specific Latent Heat of Fusion.
- Investigating Specific Latent Heat of Vaporization.

Topic 3.2 - Investigating at least one gas law.
(Choose at least one from the following titles)
- Investigating Boyles' Law.
- Investigating Charles' Law.
- Investigating Lussac's Law.

Topic 4.2 - Investigating the Speed of Sound.

Topic 4.4 - Determining refractive index.
(Choose at least one from the following titles)
- Determining the Refractive Index of Glass by Real and Apparent Depth
- Investigating Refraction of Light, Refractive Index and Critical Angle

Topic 5.2 - Investigating factors that affect resistance.

Topic 5.3 - Determining Internal Resistance.

Topic 7.1 - Investigating Half-Life.

Topic 9.3 - Investigating Young's Double-Slit.

Topic 11.2 - Investigating a diode bridge rectification circuit.

Option C.2 - Investigating the Optical Compound Microscope.

Option C.2 - Investigating the Performance of a Simple Optical Astronomical Refracting Telescope.

DETERMINING THE ACCELERATION OF FREE-FALL

Aim:
This experiment is designed to introduce you to the detail that is needed when constructing and presenting your table of results for an experiment. It will also allow you to practice some of the data analysis techniques that have been recently discussed in class.

Procedure:
1. For various heights, drop the object as it falls to the floor, accelerating at a rate of g = 10m/s².
2. Time the object 3 times as it falls and take an average of these times.

Theory:
The relationship between displacement, time, initial velocity and acceleration due to gravity for an object, falling close to the Earth's surface is given by:

$$s = ut + \tfrac{1}{2}gt^2$$

IB Criteria Assessed

Criteria assessed	Level awarded
PE	/2
EX	/6
A	/6
EV	/6
C	/4

Analysis:
- Collect and record pairs of data (height and time) including units and uncertainties.
- Present these data clearly in a suitable table.
- Process your raw data in a way which will allow you to accurately calculate the value of g (acceleration due to gravity) – HINT Use the theory above to manipulate your data to give a linear relationship
- Take into account any errors or uncertainties in your processed data.
- Draw a suitable graph that will allow an analysis to be made on the raw data.
- From your graph calculate the acceleration due to gravity, g, for a falling object.
- Include actual and percentage uncertainties by plotting suitable straight lines to determine these errors.
- Make valid conclusions related to the value calculated for g found. Compare this calculated value of g to literature values.

Evaluation:
- Evaluate the procedure, including any modifications you had to make to overcome problems
- Include an evaluation of the apparatus used.
- Suggest ways in which the procedure (and apparatus) could be modified in order to improve future investigations.

DETERMINING SPECIFIC HEAT CAPACITY BY THE ELECTRICAL METHOD

Aim:
1. To calculate the SHC of aluminum, copper, iron and water.
2. To see how accurately you can measure a known quantity.
3. To appreciate the effect of heat loss - this is important in all heat experiments.

IB Criteria Assessed

Criteria assessed	Level awarded
PE	/2
EX	/6
A	/6
EV	/6
C	/4

Procedure:
1. Setup the circuit below.
2. Take the initial temperature of the material being tested.
3. Turn on the power supply and set it to 12V. Start the stop watch.
4. Record the temperature, voltage, current at regular time intervals and record them.
5. Continue to record data for 20 minutes.
6. Repeat the above procedure for the other two materials under test

Diagram:

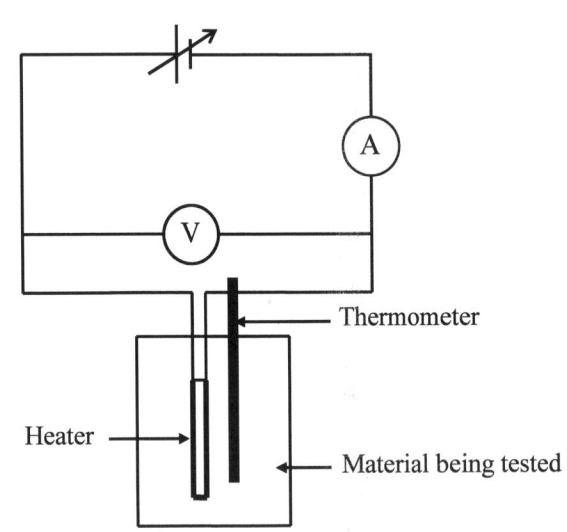

Apparatus:
Aluminum block,
Copper block,
Iron block,
Beaker of water,
Power supply,
Thermometer,
Ammeter,
Voltmeter,
Connecting leads,
Stop watch,
Heater,
Digital balance.

Theory:

| Energy supplied by heater | = | Energy received by aluminum block |
| V I t | = | m c ΔT |

Analysis:
- Record temperature, voltage, current and time in a suitable table.
- Your results table and the presentation of data should include any uncertainties associated with the apparatus that you have used.
- Draw a suitable graph that will allow you to use the formula above to calculate the specific heat capacities of the three materials.
- Your conclusion should include values for the specific heat capacities of the 3 different materials and a comparison with literature values.

Evaluation:
- Evaluate the procedure including any modifications you had to make to overcome problems. Include an evaluation of the apparatus used.
- Suggest ways in which the procedure could be modified in order to improve it for the future

DETERMINING SPECIFIC HEAT CAPACITY BY THE METHOD OF MIXTURES

Aim:
1. To calculate the SHC of aluminum, copper, iron and water
2. To see how accurately you can measure a known quantity.
3. To appreciate the effect of heat loss - which is important in all heat experiments.

Procedure:
1. Measure the mass of the material under test
2. Heat the water and test material in a beaker and keep boiling for about 5 minutes (to ensure the test material is at 100°C)
3. Measure the initial temperature and mass of the cold water in the other beaker
4. Move the test material from the hot water to the cold water
5. Stir the water
6. Wait until the water, test material and beaker reach thermal equilibrium
7. Take the final temperature of the mixture

IB Criteria Assessed

Criteria assessed	Level awarded
PE	/2
EX	/6
A	/6
EV	/6
C	/4

Diagram:

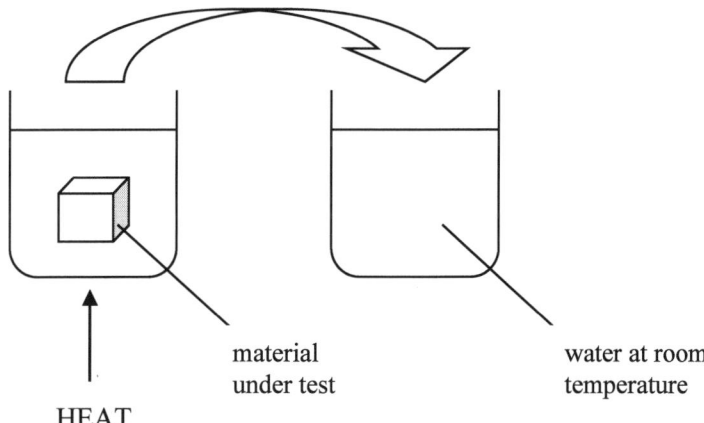

material under test

HEAT

water at room temperature

Apparatus:
aluminum block
copper block
iron block
2 beakers of water
thermometer
Bunsen burner / heater
balance

Theory:

> **Energy supplied by test material   =   Energy received by water and beaker**

Analysis:
- Record your results in a suitable table.
- Your results table and the presentation of data should include any uncertainties associated with the apparatus that you have used.
- Calculate the specific heat capacity of the three test materials.
- Your conclusion should include values for the specific heat capacities of the 3 different materials and a comparison with literature values.

Evaluation:
- Evaluate the procedure including any modifications you had to make to overcome problems. Include an evaluation of the apparatus used.
- Suggest ways in which the procedure could be modified in order to improve it for the future.

Investigating Specific Latent Heat of Vaporization of Water

Aim:
This experiment is to show you that we can do scientific experiments with non-specialized equipment (in this case an ordinary kettle) and still obtain reasonable results. You will be able to compare your result with the accepted literature value. Also, this experiment is useful as one that can be done in a very short time to obtain an answer that is fairly close to the true value - which is a useful skill in science.

Apparatus:
Electric kettle, scales, stop watch.

IB Criteria Assessed

Criteria assessed	Level awarded
PE	/2
EX	/6
A	/6
EV	/6
C	/4

Diagram:

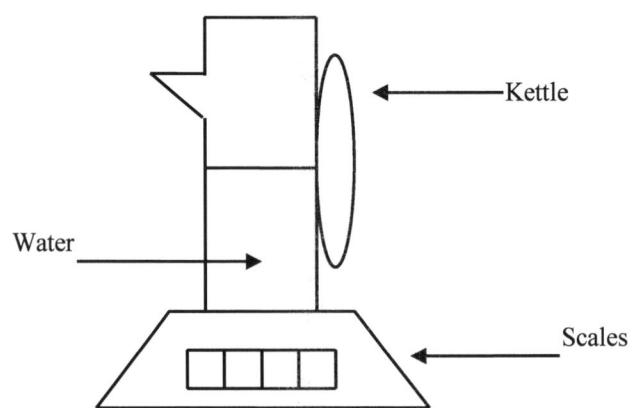

Procedure:
Put the kettle, ½ full of water on the scales. Switch it on and wait for the water to start to boil. When it is clearly boiling, start the stop watch and record the mass of the kettle every 30 seconds. Continue to take mass measurements until the mass of the kettle has fallen by 50 g. Repeat the above procedure a number of times to minimize uncertainties.

Theory:

> Energy supplied by the kettle = Energy received by the water

From the above equation you can calculate the specific latent heat of vaporization of water (the power of the kettle is written on it).

Analysis:
- Record all your data in a suitable table of results.
- Record the uncertainties in your measurements
- Using the equation given, plot a suitable graph that will allow you to calculate the Specific Latent Heat of Vaporization of Water.
- Include an estimate of the uncertainty in your calculated value
- Give a conclusion and explanation of your results; compare to literature values.

Evaluation:
- Evaluate the above procedure and apparatus used, including limitations, weaknesses or errors.
- Identify any weaknesses and suggest ways of improving the investigation.

INVESTIGATING SPECIFIC LATENT HEAT OF FUSION OF ICE

Aim:
All heat experiments have problems with heat loss or gain from the surroundings. This experiment contains a trick to try and get round this difficulty. The water is pre - heated 5 ^{0}C above room temperature and then cooled to 5 ^{0}C below room temperature. In this way, during the first half of the experiment the water is losing heat and in the second half it is gaining the same quantity of heat - the loss and gain of heat cancel each other out. Tricks like this are very important in experiments in order to overcome practical difficulties.

IB Criteria Assessed

Criteria assessed	Level awarded
PE	/2
EX	/6
A	/6
EV	/6
C	/4

Procedure:
Measure room temperature. Weigh the empty beaker. Add water to the beaker and weigh them again. Warm the water to 5 ^{0}C above room temperature. Add small pieces of crushed ice, stirring all the time, until the temperature falls to 5 ^{0}C below room temperature. Weigh the beaker and ice to determine the mass of ice added. Repeat the experiment to get 2 sets of results.

Diagram:

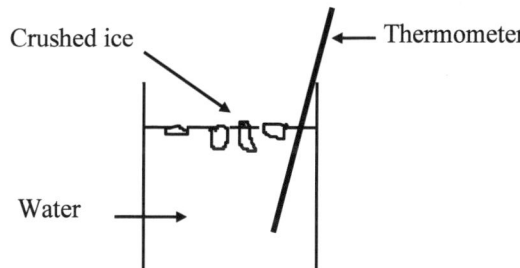

Apparatus:
Beaker
thermometer
crushed ice
scales
Bunsen burner or kettle.

Theory:

> Energy received by ice = Energy released by water

From this equation you can calculate the specific latent heat of fusion of ice.

Analysis:
- Record temperature, mass of water and mass of ice in a suitable table.
- Your results table and the presentation of data should include any uncertainties associated with the apparatus that you have used.
- Process your date in such a way that the Specific Latent Heat of Fusion for the ice can be established

Evaluation:
- Your evaluation should include values for the specific latent heat of fusion for ice and a comparison with literature values.
- Were your 2 results very different?
- You should also comment on the procedures and apparatus used, and suggest possible causes of errors and modifications that could be introduced to improve the investigation.

Original Lab Sheet by Brian Seve – Modified to the current IB syllabus requirements by Mike Dickinson

INVESTIGATING BOYLE'S LAW

Aim:
The pressure, volume and temperature of a fixed mass of gas are related to each other, but in order to study the effects we must first fix one of the variables (pressure volume or temperature) and see how the other two change. In the case of Boyle's Law our aim is to investigate the relationship between pressure and volume so we will keep the temperature constant.

IB Criteria Assessed

Criteria assessed	Level awarded
PE	/2
EX	/6
A	/6
EV	/6
C	/4

Diagram:

Apparatus:
Gas syringe
Pressure sensor
Computer
Data-logging software

Theory:
For a fixed mass of gas, the volume is inversely proportional to the pressure, if the temperature is kept constant. This is a statement of Boyle's law. If P_1 and P_2 are the initial and final pressures and V_1 and V_2 are the initial and final volumes, then Boyle's law may be expressed as:

$$P_1 V_1 = P_2 V_2$$

Procedure:
Connect the gas syringe to the pressure sensor ensuring an air-tight connection. Connect the pressure sensor to the computer and start the data logging software. Apply a force to the syringe's plunger so that the volume in the syringe is reduced. Check that a corresponding change occurs in the pressure reading on the computer.

Analysis:
- Collect of pairs of results for P and V.
- Present these data (together with uncertainties) in an appropriate results table.
- Plot a suitable graph from which you can deduce a relationship between pressure and volume.
- If the graph is non-linear, turn it into one which is.
- Do your graphs verify the theory above? Explain

Evaluation:
- Evaluate the data that you have collected and analyzed - compare your results to the theory and explain any discrepancies
- Evaluate the procedure, including any modifications you had to make to overcome problems. Include an evaluation of the apparatus used.
- Suggest ways in which the procedure could be modified in order to improve it for the future.

INVESTIGATING CHARLES' LAW AND ABSOLUTE ZERO

Aim:
Physics has many hundreds of laws and these can be tested in experiments to see if they are true. In 1787 the French scientist, J.A.C. Charles, published a law connecting the volume and temperature of gases. Your task is to see if you agree with his law. Another thing you have to consider is "is my experiment accurate enough to prove or disprove the law?" It could be that the law is good but your experiment isn't. This process of testing a proposed law with an experiment is one of the foundations of all science.

IB Criteria Assessed

Criteria assessed	Level awarded
PE	/2
EX	/6
A	/6
EV	/6
C	/4

Apparatus:
Beaker, thermometer, capillary tube with plug of concentrated sulfuric acid, 30 cm ruler, Bunsen burner.

Diagram:

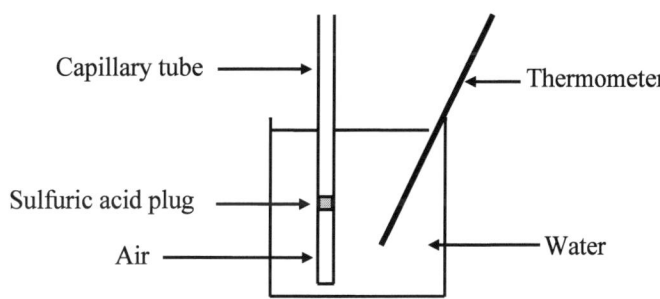

Procedure:
Put the capillary tube in the beaker as shown. Measure the length of the air below the plug in the capillary with a ruler. <u>Slowly</u> warm the water and approximately every 10 °C, measure the length of air again. Repeat this until the water is boiling. Make sure that the air column in the capillary tube is always below the level of the water.

Analysis:
- Collect pairs of data for length of air and temperature
- Plot a graph of "length of air" against "temperature in °C".
- Extrapolate the line until it reaches the x axis (y=0)
- Give a conclusion and explanation of your results; compare your results to literature values.

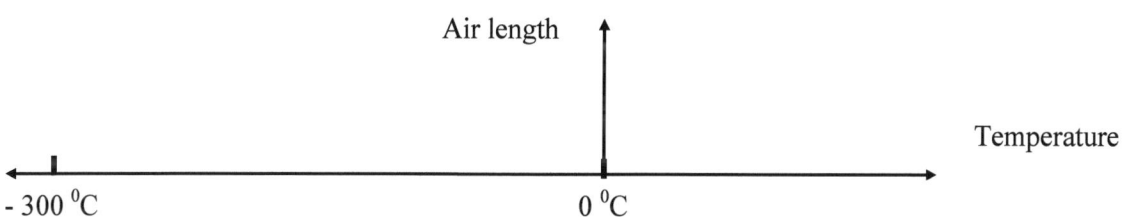

Evaluation:
- Evaluate the above procedure and apparatus used, including limitations, weaknesses or errors.
- Identify any weaknesses and suggest ways of improving the investigation.

INVESTIGATING LUSSAC'S LAW
(THE PRESSURE LAW)

Aim:
The early gas laws were developed at the end of the 18th century, when scientists began to realize that relationships between the pressure, volume and temperature of a sample of gas could be obtained which would hold for all gases. Gay-Lussac's law, or the pressure law, was found by Joseph Louis Gay-Lussac in 1809. It states that the pressure exerted on the sides of a container by an ideal gas of fixed volume is proportional to its temperature [Wikipedia.org]. Your task is to see if you agree with his law. Another thing you have to consider is "is my experiment accurate enough to prove or disprove the law?" It could be that the law is good but your experiment isn't. This process of testing a proposed law with an experiment is one of the foundations of all science.

IB Criteria Assessed

Criteria assessed	Level awarded
PE	/2
EX	/6
A	/6
EV	/6
C	/4

Diagram:

- Thermometer
- Clamp and stand
- Fixed Volume of air
- Water
- HEAT
- Bourdon Pressure Gauge

Apparatus:
Clamp-stand and clamp
Thermometer
(or temperature sensor)
Bunsen burner
(or electric heater)
Tripod
Heatproof mat
Gauze
Large beaker
Round-bottom flask
Rubber bung
Bourdon pressure gauge
(or pressure sensor)

Procedure:
Slowly warm the water such that the air in the flash is gently heated, indirectly as shown. Turn off the heat at every 10°C to allow equilibrium to occur (remember you are measuring the temperature of the water and assuming that this is the same temperature as the air in the flask). Repeat this until the water is boiling.

Analysis:
- Collect pairs of data for pressure and temperature.
- Plot a graph of "pressure" against "temperature in °C". Take account of atmospheric pressure if necessary.
- Extrapolate the line until it reaches the x axis (y=0)
- Give a conclusion and explanation of your results; compare your results to literature values.

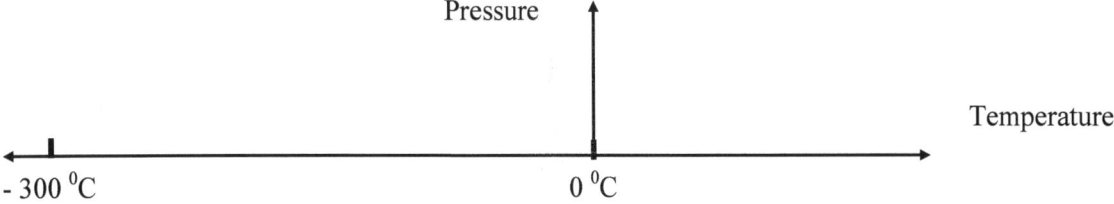

Evaluation:
- Evaluate the above procedure and apparatus used, including limitations, weaknesses or errors.
- Identify any weaknesses and suggest ways of improving the investigation.

INVESTIGATING RESONANCE
DETERMINING THE SPEED OF SOUND

Aim:
Sound travels at approximately 330 m/s. This is too fast to measure with a stop watch. A clever way of measuring the speed of any wave (sound, light, radio etc.) is to freeze it. With a stationary wave, since it is not moving, it is easy to measure the wavelength and, if you also know the frequency of the wave, you can calculate the speed of the wave. This is yet another example of the use of a clever "trick" in order to measure something which is very difficult to measure directly.

IB Criteria Assessed

Criteria assessed	Level awarded
PE	/2
EX	/6
A	/6
EV	/6
C	/4

Apparatus:
Large measuring cylinder, glass tube, set of tuning forks, meter rule.

Diagram:

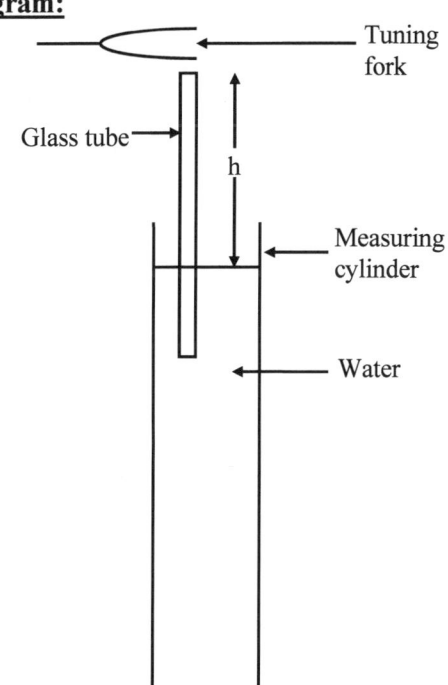

Procedure:
1. Hold the vibrating tuning fork above the glass tube.
2. Move the tube up and down until you find the <u>shortest</u> length of the air column that produces a loud sound.
3. Measure length "h". Note the frequency of the tuning fork.
4. Repeat this for tuning forks of several different frequencies.

Theory:
You have found the fundamental resonance length for the different frequencies. The length at which resonance occurs for the different frequencies is a function of the wavelength of the sound wave in the glass tube. The relationship between the frequency, wavelength and speed of the wave is given by;

$$v = f\lambda$$

Analysis:
- Collect of pairs of results for f and h for when the sound from the air column is at its loudest.
- Present these data (together with uncertainties) in an appropriate table.
- Plot a suitable graph from which you can calculate the speed of sound.
- If your graph does not go through the origin, then there may be a reason for this. (hint - look in your text book for "end correction")

Evaluation:
- Evaluate the data that you have collected and analyzed. Compare the speed of sound obtained experimentally with the value published in your text-book.
- Evaluate the procedure, including any modifications you had to make to overcome problems. Include an evaluation the apparatus used.
- Suggest ways in which the procedure could be modified in order to improve it for the future.

Determining the Refractive Index of Glass by Real and Apparent Depth

Aim:
When you look down at your feet in a swimming pool, they appear to be closer than normal. This is a short experiment that uses this phenomenon to produce a reasonably accurate result for the refractive index of glass. So you have taken an everyday observation and invented an experiment to obtain a quantitative measurement of what you saw and calculate the uncertainty in your final answer.

Apparatus:
Travelling microscope, 2 glass blocks, lycopodium powder.

IB Criteria Assessed

Criteria assessed	Level awarded
PE	/2
EX	/6
A	/6
EV	/6
C	/4

Diagram:

(Diagram showing travelling microscope with eye piece focused on powder placed between glass blocks)

Procedure:
Put a small amount of powder on the top of the first block and focus the microscope on it - take reading d_1. Place the second block on top of the first and focus on the powder again - take reading d_2. Put a small amount of powder on top of the second block and focus the microscope on it - take reading d_3.

Analysis:
- Record the data, as specified in the above procedure, in a suitable form
- Include units and uncertainties.
- Use your raw data to accurately calculate a value for the refractive index of the glass.
- Include the uncertainty in your final value and show your calculations.

INVESTIGATING REFRACTION OF LIGHT, REFRACTIVE INDEX AND CRITICAL ANGLE

Aim:
1. To investigate the relationship between the angles of incidence and refraction as light travels into a rectangular Perspex block.
2. To verify experimentally the formula: Refractive index n = sin i / sin r
3. To investigate total internal reflection and the "critical angle", C, of a refractive substance.
4. To verify experimentally the formula:
5. Refractive index n = 1 / sin C

IB Criteria Assessed

Criteria assessed	Level awarded
PE	/2
EX	/6
A	/6
EV	/6
C	/4

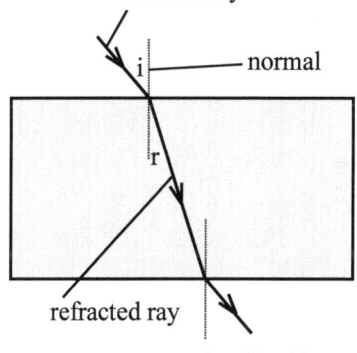

r = angle of incidence
i = angle of incidence

Procedure (Part A):
1. Place the rectangular Perspex block on a sheet of white paper.
2. Draw around the block (in case it gets disturbed).
3. Shine a single, thin ray of light from a ray box, incident to one long face of the block.
4. Mark the incident ray entering the block and the emergent ray exiting the block.
5. Measure the angle of incidence (i) and the corresponding angle of refraction (r).
6. Vary the angle of incidence so that 8 different pairs of results can be collected.

Procedure (Part B):
1. Place the semi-circular Perspex block on a sheet of white paper.
2. Draw around the block (in case it gets disturbed).
3. Shine a single, thin ray of light from a ray box, incident to the curved face of the block, pointing at the center of the straight face
4. Mark the incident ray entering the block and the emergent ray exiting the block.
5. Increase the angle of incidence (i) until the emergent ray disappears and the light is reflected inside of the glass block.
6. Measure angle i and the corresponding angle of <u>reflection.</u>

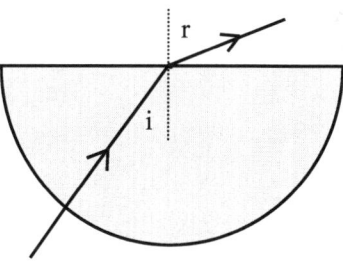

Analysis:
- From Part A, collect and record pairs of data (angles i and r) including units and uncertainties.
- Present these data clearly in a suitable table.
- From Part B, record the critical angle C.
- Use your results from part A of the experiment to plot a suitable graph that will verify the formula for refractive index of Perspex.
- Calculate the refractive index of the Perspex using your graph.
- Use the results from part B to calculate the refractive index of the Perspex.
- Compare the results for refractive index from the two experiments.
- Which one do you think is the most accurate? Why?

INVESTIGATING FACTORS THAT AFFECT RESISTANCE

Introduction:
Copper and other good electrical conductors allow "free" (conduction) electrons to pass through them very easily. The molecular arrangement in a piece of Nichrome wire on the other hand disrupts this free flow of electrons and provides some resistance to their movement. There are many factors which determines the amount of resistance that the piece of wire provides.

Aim:
You are to investigate a factor that affects the resistance of a piece of Nichrome wire.

Apparatus:
One meter of Nichrome wire, plus any other materials you may need.

Practice Exploration:
Design a procedure that will allow you to investigate a factor (or factors) that affect the resistance of a piece of wire. This procedure should include the following sections:
- Defining the Problem and selecting variables:
- Controlling the Variables:
- Developing a procedure for collecting data:

Include a quantitative hypothesis for your investigation.

Analysis:
- Show the results for the experiment in a suitable table. Include uncertainties.
- Use suitable graphs to allow for a full analysis to be carried out on your chosen variable(s)

Evaluation:
- Evaluate the data that you have collected and analyzed. Make a suitable conclusion. Compare the result obtained from your experiment to literature values (if possible)
- Evaluate your own plan, including any modifications you had to make to overcome problems. Include an evaluation of the apparatus used.
- Suggest ways in which the procedure could be modified in order to improve it for the future.

IB Criteria Assessed

Criteria assessed	Level awarded
PE	/2
EX	/6
A	/6
EV	/6
C	/4

Determining the EMF and Internal Resistance of a Cell

Aim:
To determine the emf and internal resistance of a cell (battery) using a graphical method.

Procedure:
1. Set up the circuit as shown in the diagram with the variable resistor set to its maximum value.
2. Before closing the switch, record the reading, V, on the voltmeter
3. Close the switch
4. By adjusting the variable resistor, obtain pairs of voltmeter and ammeter readings over the widest possible range.
5. Open the switch after each pair of readings and only close it for as long as is necessary to obtain each pair of readings.

IB Criteria Assessed

Criteria assessed	Level awarded
PE	/2
EX	/6
A	/6
EV	/6
C	/4

Apparatus:
Dry cell (battery), ammeter, voltmeter, variable resistor, switch.

Diagram:

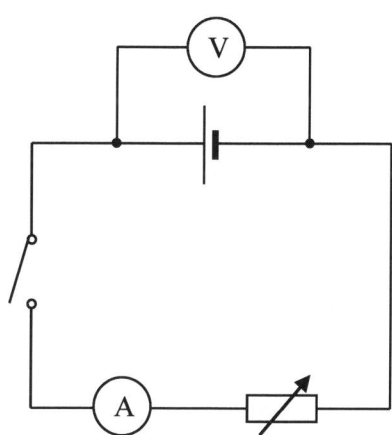

Analysis:
- Collect and record pairs of results for Voltage and Current including units and uncertainties.
- Present these data clearly.
- Process your raw data in such a way that will allow you to accurately calculate the value of emf and internal resistance for the cell.
- Take into account any errors or uncertainties in your processed data.
- Give a conclusion and explanation of your results; compare to literature values if possible.

Evaluation:
- Evaluate the above procedure and apparatus used, including limitations, weaknesses or errors.
- Identify any weaknesses and suggest ways of improving the investigation.

INVESTIGATING RADIOACTIVE DECAY AND HALF-LIFE (SIMULATION USING DICE)

Aim:
This experiment simulates radioactive decay. Instead of nuclei decaying, we will be looking at dice with the number "1" showing. These dice will represent decayed atoms and will be removed at each step.

Apparatus:
100 dice, large container.

Procedure:
1. Put the 100 dice into a container.
2. Shake the container and pour the dice onto the desk.
3. Remove decayed nuclei and record the number involved.
4. Put the remaining dice back into the container and repeat steps 2 and 3 several times.
5. Repeat, to get two full sets of data.
6. Share your results with the other teams so that you can calculate an average value for the "number of remaining dice" for each roll.

IB Criteria Assessed

Criteria assessed	Level awarded
PE	/2
EX	/6
A	/6
EV	/6
C	/4

Analysis: (BOTH Standard and Higher level)
- Record your results for the simulation in a suitable table.
- Your results table and the presentation of data should include any uncertainties associated with the apparatus that you have used.
- Draw a suitable graph that will allow you to calculate the half-life of your decay model.
- Show your working clearly and explain your method used.
- State your final value for the half-life for your decay model
- Explain if the background count rate was included in this graph.

Analysis: (Higher level only)
- The radioactive equation can be written:

$$N = N_O e^{-\lambda t}$$

- Rearrange this formula so that it is of the form y = mx + c
- Draw a suitable log graph that will allow you to obtain a value for the decay constant, λ and calculate the half-life, t, of your decay model.
- Compare the two values for half-life obtained from the graphs.
- Which method do you think is more accurate? State your reasons.

Determining the Wavelength of Laser Light using Young's Double Slits

Aim:
This experiment is historically very important because with it Thomas Young showed that light is a wave and so finished off Newton's "corpuscular theory" of light. If light is a wave then it should produce interference and also it should be possible to use the interference pattern to measure the wavelength of the light - this is exactly what Young was able to do - therefore proving that light is a wave.

IB Criteria Assessed

Criteria assessed	Level awarded
PE	/2
EX	/6
A	/6
EV	/6
C	/4

Procedure:
Put the double slits in a stand and clamp as far as possible from the wall or screen. Aim the laser beam at the slits. Measure the distance from the slits to the wall and also the distance between the dots on the wall. Use the Vernier to measure the double slit separation.

DO NOT SHINE THE LASER LIGHT IN YOUR EYES - IT COULD BLIND YOU ! !

Apparatus:
Laser, Vernier gauge, meter rule, double slits, clamp and stand.

Diagram:

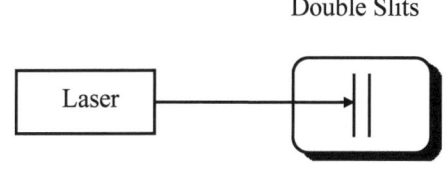
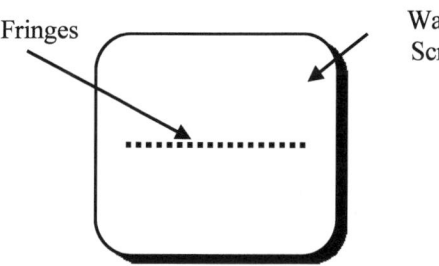

Theory:
Use the equation:

$$\text{Wavelength} = \frac{\text{Slit separation} \times \text{Fringe width}}{\text{Distance from slits to screen}}$$

Analysis:
- Show the results for the experiment in a suitable table. Include uncertainties.
- Calculate a value for the wavelength of laser light.
- Calculate the actual and percentage errors in this value, based on the uncertainties in the apparatus that you have used.

Evaluation:
- Evaluate the data that you have collected and analyzed. The correct value for He / Ne laser light is given in the data book. Try to identify the errors in your experiment that caused the difference between your result and the true result.
- Evaluate the procedure, including any modifications you had to make to overcome problems. Include an evaluation the apparatus used.
- Suggest ways in which the procedure could be modified in order to improve it for the future.

INVESTIGATING A DIODE BRIDGE RECTIFICATION CIRCUIT

Aim:
Rectification is the process by which an alternating current is turned into direct current by the cleaver use of (usually) four diodes. In this experiment you will become familiar with the diode bridge rectifier and to investigate various methods of passive filtering.

IB Criteria Assessed	
Criteria assessed	Level awarded
PE	/2
EX	/6
A	/6
EV	/6
C	/4

Diagram:

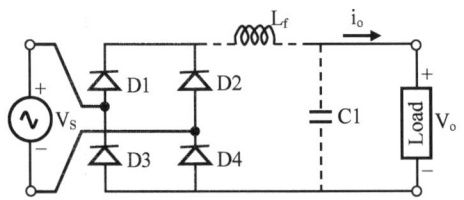

Theory:
Both of the diagrams above show the same single-phase diode bridge rectifier circuit using two different styles of diagram. The one on the left is the style of diagram as might be used in a textbook, where the one on the right uses a style that allows students to more readily build the circuit using a breadboard. To understand the single-phase full-bridge operation, refer to the diagram above and consider the case where the source voltage, V_s goes positive. For the moment, assume no filter components (L and C) are present and the load is purely resistive. It makes sense that diode, D1, will tend to turn on (since its anode is going positive). At the same time, diode, D4, will tend to turn on (since its cathode is going negative). Thus, with both D1 and D4 on, the output voltage is positive. During the next half cycle, i.e., the ac source goes negative, diodes D2 and D3 now turn on simultaneously, and again the output voltage is positive. In this way, the current that flows into the load is always positive.

Apparatus:
Signal generator, diodes (e.g. 1N4001), capacitors (15μF, 400μF and 4,700μF work well), resistors (e.g. 22kΩ) and inductors (0.6H or thereabouts), breadboard and oscilloscope

Procedure:
1. Connect the oscilloscope directly to the signal generator (fig. 1) and adjust the supply voltage, V_s from the signal generator until a sinusoidal waveform is shown that reads a 10V peak-to-peak voltage (V_s = 5V a.c.)
2. Record the output signal on the oscilloscope in your notebook.
3. Using the breadboard and the apparatus supplied, build the circuits shown in figures 2, 3 and 4.
4. Record the output signal for each of the circuits that you have built.

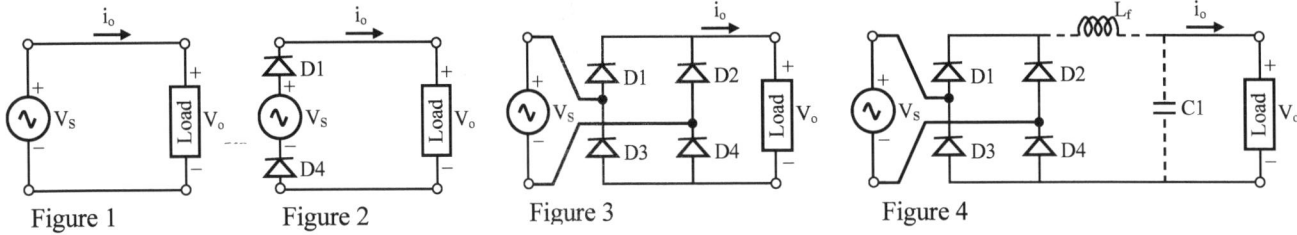

Figure 1   Figure 2   Figure 3   Figure 4

5. Introduce a capacitor and an inductor to "smooth" the double-peak waveform that you recorded from the circuit diagram in figure 3. I.e. build the circuit in figure 4.
6. Investigate various positions and values for these smoothing components.

Analysis:
- Use the data that you have collected to explain how the single-phase diode bridge rectifier circuit works
- Explain the necessity of the "load" resistor
- Why is the capacitor and inductor included in the circuit?

INVESTIGATING THE COMPOUND MICROSCOPE

Aim:
To build a microscope using two converging lenses

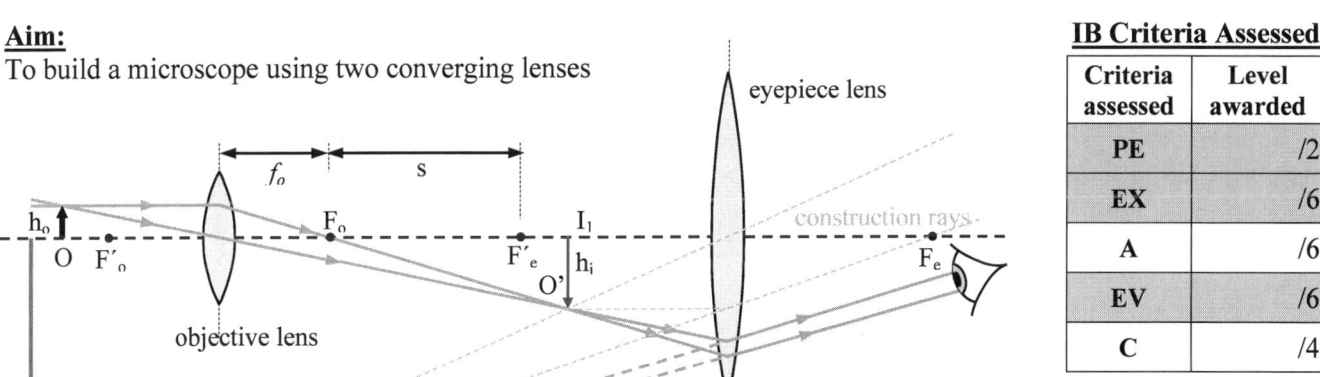

IB Criteria Assessed	
Criteria assessed	Level awarded
PE	/2
EX	/6
A	/6
EV	/6
C	/4

In a compound microscope, a small object, O is placed near, but just beyond, the focal point of the objective lens so that an enlarged, real, inverted image I_1 is formed, which becomes a real object, O' for the eyepiece lens. The eyepiece then acts as a simple magnifier to produce a final image I' which remains inverted and in the figure above. If s is the distance between the focal points F_o and F'_e, while h_o and h_i are the heights of the object and image produced by the objective, then the lateral magnification of the objective is based on similar triangles.

$$m_{obj} = \frac{h_i}{h_o} = \frac{s}{f_o}$$

The total magnification of the microscope is the product of the lateral magnification m_{obj} of the objective and the angular magnification $m_{eye} = n/f_e$ of the eyepiece; where, L is the distance between the lenses, and n is the near-point (conventionally taken to be 25cm.

$$M = m_{obj} m_{eye} = \frac{s}{f_o} \frac{n}{f_e} = \frac{(L - f_o - f_e)n}{f_o f_e}$$

Procedure:
Mount a ruled translucent scale a few centimeters beyond the (secondary) focal point of the objective lens, to serve as an object to examine (the objective should be the strongest lens). Measure the distance o between this object and the objective. Then use the thin lens equation:

$$\frac{1}{o} + \frac{1}{i} = \frac{1}{f}$$

to compute the distance i from the objective to the image produced by it. Now position the eyepiece at a distance from the objective equal to $i + f_e$. Finally, bring your eye up close to the eyepiece and adjust the position of the ruled scale until you see a focused image of it. Note that the microscope will invert the image relative to the object.

To determine the experimental magnification of the microscope
Look at an adjacent pair of rulings on the object screen through the microscope, back your eye up slightly, and use a felt pen to mark the apparent positions of the two rulings directly on the eyepiece lens. Now measure the separation between your two marks with a ruler. Also measure the actual separation between the two rulings on the screen. The ratio of these two separations is the magnification.

To determine the theoretical magnification of the microscope
Measure your near point distance; hold up a plastic ruler to the bridge of your nose with the zero end at your nose. Gradually bring a small object which has writing on it along the ruler toward your eye (with your glasses on). Find the shortest distance at which you can still comfortably focus on the writing (i.e., without having to strain so much that you get a headache: be reasonable!). Use your near point distance n to compute the theoretical magnification as $M = (L - fo - fe) \times n /(fo \times fe)$ where L is the measured distance between the two lenses.

Compare the theoretical magnification to the experimental magnification.

Original Lab idea from the University of West Florida – Modified to the current IB syllabus requirements by Mike Dickinson

INVESTIGATING A SIMPLE OPTICAL ASTRONOMICAL REFRACTING TELESCOPE

Aim:
To build an astronomical telescope using two converging lenses separated by a distance equal to the sum of their focal lengths.

IB Criteria Assessed	
Criteria assessed	Level awarded
PE	/2
EX	/6
A	/6
EV	/6
C	/4

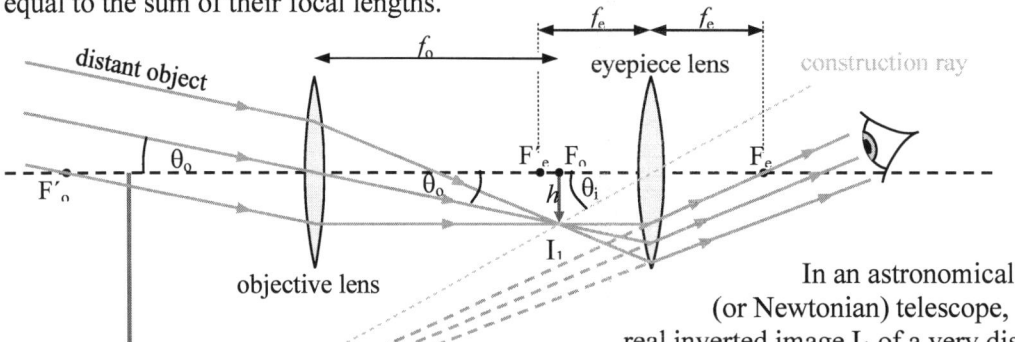

In an astronomical (or Newtonian) telescope, the objective lens forms a real inverted image I_1 of a very distant (such that $o \approx \infty$) object in its focal plane, as shown in the figure. The angular size of the image formed by the objective is given by $\theta_0 \cong \dfrac{h_i}{f_0}$ where h_i is the height of the image and f_o is the focal length of the objective. This is also the angular size of the distant object.

The image, I_1 of the objective is the real object for the eyepiece, which is used as a simple magnifier. The eyepiece then produces a virtual image, I' which is magnified in size to but remains inverted. The angular size of the image seen through the eyepiece lens is seen to be $\theta_e \cong \dfrac{h_i}{f_e}$ where f_e is the focal length of the eyepiece. The image I formed by the objective is very near the focal point of the eyepiece, so that the distance between the two lenses is approximately equal to the sum of their focal lengths, $L = f_o + f_e$. The total angular magnification of the telescope is

$$M = \frac{\theta_e}{\theta_i} = \frac{f_i}{f_e}$$

The objective must be weaker (have a longer focal length) than the eyepiece in order to give magnification (M > 1).

Procedure:
Build an astronomical telescope using two converging lenses separated by a distance equal to the sum of their focal lengths. Use the stronger lens as the eyepiece. Bring your eye up close to the eyepiece and look through it at a distant object; it should appear inverted. If you normally wear glasses, keep your glasses on. Slightly adjust the distance between the two lenses to focus the telescope. (Note: If your eye is too far from the eyepiece, you can get fooled into thinking the various instruments constructed in this lab are focused when they actually are not.)

Analysis:
Determine the experimental magnification of your telescope by looking at two parallel lines which you have drawn on the blackboard across the room. Hold up a ruler at about arm's length (but not so far that you cannot read the scale!), look through the telescope with one eye, and measure the apparent spacing between the two lines with your other eye. (If you cross your eyes you may be able to overlap the two images, making this task easier.) Now move your head slightly so that you are looking at the two lines with your naked eye, and again measure the apparent spacing between the two lines without changing the distance from the ruler to your eye. The ratio of the spacings measured with and without the telescope is the experimental magnification. Compare the result to the theoretical magnification which equals the absolute value of the ratio of the two focal lengths of the lenses. (Accuracies are not really good enough to warrant a calculation of percent error. Instead discuss in your report whether the agreement seems reasonable and why the accuracy isn't expected to be terribly high.) What happens if you turn the telescope around and look through the wrong end of it? Explain why this happened in a report.

Other Suggested Practical Activities

INVESTIGATING MURPHY'S LAW

Aim:
Murphy's Law states that if there are a number of possible outcomes to a situation, the actual outcome will be the worst possible one! One application of this law is breakfast. If you drop your toast, it will always land jam side down.

Apparatus:
Anything you need to perform the experiment to your plan.

Practice Exploration:
Design a procedure to test Murphy's Law that includes appropriate use of apparatus for the control, collection and analysis of data. You don't have to use toast if you don't want to – you can test Murphy's Law (that if it can go wrong, it will go wrong) using any scenario of your choice.

This procedure should include the following sections.

- Defining the Problem and selecting variables:
- Controlling the Variables:
- Developing a procedure for collecting data:

Include a quantitative hypothesis for your investigation.

IB Criteria Assessed

Criteria assessed	Level awarded
PE	/2
EX	/6
A	/6
EV	/6
C	/4

Analysis:
- Record the raw data (both quantitative and qualitative) for the experiment in a suitable form. Include uncertainties due to the precision of the measuring apparatus.
- Process your quantitative raw data.
- Present the processed data in an appropriate way and include errors and uncertainties
- Draw conclusions based on your interpretation of the data

Evaluation:
- Evaluate your own plan, including any weaknesses and/or limitations. Include an evaluation of the apparatus used.
- In light of the weaknesses and limitations suggested above, suggest ways in which the procedure could be modified in order to improve it for the future.

Physics Practical Scheme of Work – For use with the IB Diploma Programme – First Assessment 2016

INVESTIGATING ERRORS AND UNCERTAINTIES IN EXPERIMENTS

Aim:
All experiments are done with measuring instruments. No instrument is perfectly accurate - they all have limits to their accuracy. It is important that you realize that no experiments give perfect, exact answers. To illustrate this you are going to measure the density of a slide with:

1. a precision electronic balance, and a Vernier gauge.
2. then with a lever arm balance and a meter rule.

From these 2 sets of measurements you will calculate the density of the slide. One of the experiments is more accurate than the other, but both are imprecise.

From now on, all your experimental results will include the uncertainties associated with the measuring apparatus used. (and for HL student, a calculation of $\pm$ error in the readings and in the final result).

IB Criteria Assessed

Criteria assessed	Level awarded
PE	/2
EX	/6
A	/6
EV	/6
C	/4

Apparatus:
Vernier calipers, meter rule, microscope slide, electronic scale, lever arm balance.

Procedure:
1. Measure the mass of the slide on both the precise and less precise scale.
2. Measure the length, breadth and thickness of the slide with the vernier and then the meter rule.
3. Write all of your measurements in a suitable table of results with a $\pm$ uncertainty at the top of each column.

Theory:

$$\text{density} = \frac{\text{mass}}{\text{volume}}$$

Calculate density using the readings from the meter rule and less precise balance. Repeat this with the readings from the precise scale and the vernier calipers. Compare the precision of these two results.

Analysis:
Calculate the maximum and the minimum possible value for the density of the slide based upon the precision of the two sets of apparatus.

Using the uncertainty in the readings, calculate the largest and smallest possible values for the density of the slide. In both cases produce an answer for the density of the slide in the form:

> Density = ***** g cm^{-3} $\pm$ ** g cm^{-3}
> Density = ***** g cm^{-3} $\pm$ ** %

All measuring instruments have limits to their accuracies but you can make them less accurate by not using them carefully. It is important that :

1. the balances are at zero before the slide is put on them.
2. you **put** the slide on the balance; don't **drop** it on them.
3. the meter rule is not worn away at one end.

In all your future experiments, you should always be aware of the importance of using your measuring instruments as accurately as possible.

INVESTIGATING UNCERTAINTIES
MEASURING INSTRUMENT CIRCUS

Aim:
You need practice in taking all measurements with a ± error and then performing calculations to arrive at a final answer in the form, for example:

$$\text{Volume} = 28.9 \text{ cm}^3 \pm 2 \text{ cm}^3$$
$$\text{or} \quad \text{Volume} = 28.9 \text{ cm}^3 \pm 6.9\%$$

There are also some useful "tricks" you can learn to improve the accuracy of your experiments. Think about these procedures and include them when collecting your data.

IB Criteria Assessed

Criteria assessed	Level awarded
PE	/2
EX	/6
A	/6
EV	/6
C	/4

Apparatus:
Meter rule, Vernier calipers, micrometer, marble, bag of rice, digital scales, mechanical balance, short length of copper wire, 20 sheets of paper, stop watch, ping pong ball, measuring cylinder, tea cup.

Procedure:
You must use the available instruments to measure the following items:
1. the diameter of the marble.
2. the surface area of the marble.
3. the volume of the marble.
4. the mass of a grain of rice.
5. the diameter of the copper wire.
6. the time for a ping pong ball to fall from a height of 3 meters.
7. the volume of liquid in a tea cup.
8. the area of the rectangle below.

Analysis:
- Record the raw data (both quantitative and qualitative) for the experiment in a suitable form. Include uncertainties due to the precision of the measuring apparatus with a ± heading for each column on your table.
- Remember that your ± error estimates are not the only sources of error in the experiments. Look for and think about other causes of errors - e.g.: is the wire the same thickness along the whole length?. There are several problems like this in the experiments. If you spot them, your experimental results will be more accurate.
- What assumptions did you make in your results? e.g. - are all rice grains the same size?
- Process your quantitative raw data.
- Specify what instrument you used and the methods you employed to improve the accuracy of your results.

DETERMINING STIFFNESS OF STEEL BY THE OSCILLATIONS OF A HACKSAW BLADE

Aim:
This is to show that a concept difficult to measure can be easily calculated indirectly from a suitable equation. Also this is further practice in plotting an appropriate graph and using the gradient to find a constant - in this case the constant is the stiffness (E) of the steel. Also you will see how to use your graph to obtain a ± estimate of accuracy.

IB Criteria Assessed

Criteria assessed	Level awarded
PE	/2
EX	/6
A	/6
EV	/6
C	/4

Apparatus:
Clamp, hacksaw blade, stop watch, 2 magnets, ruler, micrometer, digital scales, 2 blocks of wood.

Diagram:

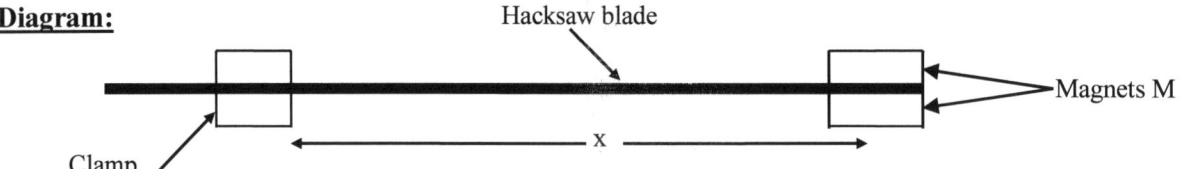

Procedure:
Clamp the hacksaw blade between the 2 blocks of wood with the blade <u>vertical</u>. Attach the 2 magnets to the <u>end</u> of the blade, about 0.15 m from the blocks (distance x). Pull the magnets to one side and release them so that the blade vibrates. Time the period T. Repeat this for various lengths of x up to 0.3 m. Also measure the mass of the magnets (M), the thickness of the hacksaw blade (d), and the breadth (b).

Analysis:
Period T, mass M and stiffness E are related by the equation:

$$T^2 = \left[\frac{16\pi^2 M}{bd^3 E}\right] x^3$$

- Draw a suitable table to include all of the variables together with their actual uncertainties and percentage uncertainties due to the apparatus used.
- Plot a suitable graph that will allow you to find the stiffness E.
- What is the accuracy of your result? To estimate this, find the best-fit line on the graph and the worst fit lines.
- Calculate a value for E from each gradient and so find a value for the uncertainty in your result for E.

Evaluation:
- What does the term "stiffness" mean?
- What is the value of the stiffness (E) for steel?
- How does this value compare with the one that you have obtained.
- Suggest reasons for any differences.
- Suggest areas where the procedures used in this practical may have been the cause of some of these errors.
- Suggest modifications to the practical to minimize any errors and shortcomings.

INVESTIGATING THE VIBRATIONS OF A LOADED METRE RULE

Aim:
Many experiments use an equation in which one of the factors is raised to a power e.g. - Z^2, $H^{1.3}$. The equation could be for example :

$$R = k \cdot Z^2$$

To test if this equation is true we need to plot a log graph from which we can find :

     (a) the power that Z is raised to (in this case it is 2)
     (b) the value of the constant k

IB Criteria Assessed

Criteria assessed	Level awarded
PE	/2
EX	/6
A	/6
EV	/6
C	/4

This is a good example of the use of mathematics to manipulate data and from it find useful extra information from the results of our experiments.

Use the experimental results below to plot a log graph and test the relationship for a vibrating ruler :

$$T = k \cdot l^{3/2}$$

Diagram:

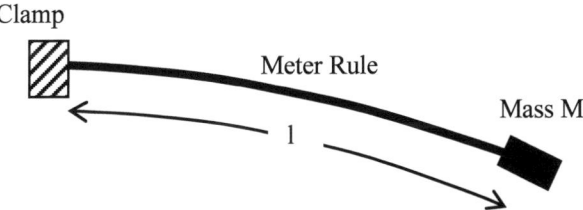

Results:
The results in the table below were obtained for the vibrations of a metre rule loaded with a mass M = 0.2 kg as the projecting length, l, was varied.

l / m	Time for 20 complete vibrations / s				
0.9	11.5	11.0	11.1	11.0	11.2
0.8	9.0	9.7	9.7	9.2	9.4
0.7	7.6	7.5	7.1	8.0	7.3
0.6	6.8	6.0	6.7	6.0	6.8
0.5	5.0	5.6	5.6	5.1	4.7

Analysis:
- Draw up a table of l and the average time for one vibration (the period), T.
- Theoretically, T = k.l3/2, where k is a constant depending on mass, M, the dimensions of the cross-section of the meter rule and the elastic properties of the material of the rule.
- In order to determine the constant k and confirm the form of the relationship in (2.), the equation has to be rewritten in log form. Include this log data in your new table.
- Plot a suitable graph in order to confirm the relationship T = k.l3/2
- From your graph, determine the value for n and k
- Comment on whether these results confirm the theoretical relationship with a value of n = 1.5. Can you suggest any reasons for deviations from the linear relationship expected in (4.)?

INVESTIGATING THE RANGE OF GAMMA RADIATION IN AIR

Aim:
The suggested relationship between the count rate for a gamma radiation source and the distance away from a Geiger-Müller tube is of the form:

$$\frac{1}{\sqrt{N_C}} = k(l + L)$$

Where l is the distance between the G-M tube window and the protective grille covering the source, N_C is the count rate **per minute** (corrected for background) and where k and L are constants.

IB Criteria Assessed

Criteria assessed	Level awarded
PE	/2
EX	/6
A	/6
EV	/6
C	/4

The aim of this investigation is to calculate the constants k and L.

Procedure:
A Geiger-Müller tube was fixed in the laboratory with its window pointing downwards, well away from all known radioactive sources. Background count-rate measurements were made by recording counts over several five minute periods. The following counts were obtained for five such periods: 128, 138, 123, 130 and 145.

With the Geiger-Müller tube in the same position a weak radioactive source, covered by a protective grille, was placed below it as shown in the diagram.

Diagram:

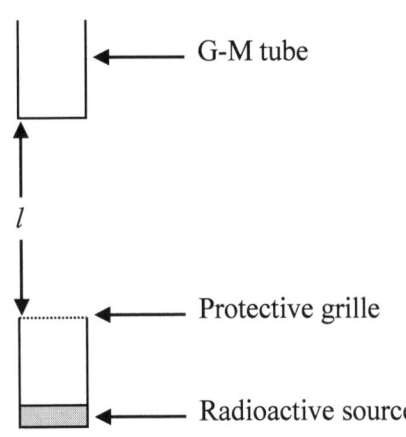

Results:
The number of counts, N, over a two minute period was noted for different distances, l. The results are recorded in the table below.

l / m	N / number of counts
15	8910
30	3990
45	2268
60	1471
75	1038
90	777
105	607

Analysis:
1. Copy the table above and complete it by including columns for N_C and $1/\sqrt{N_C}$.
2. Plot a suitable graph which will allow you to verify the theory above (aim).
3. Use your graph to calculate the constants k and L, showing all your working.
4. Comment on the validity of the suggested relationship and suggest a physical significance for L.

INVESTIGATING FORCES IN EQUILIBRIUM

Aim:
1. To use vector addition to calculate the mass of an unknown object.
2. To gain an understanding of balanced forces in situations of static equilibrium when two or more forces act at a point in a system.

Equipment:
2 pulleys mounted on a board, thread, slotted masses, 3 sheets of white paper for each student.

IB Criteria Assessed

Criteria assessed	Level awarded
PE	/2
EX	/6
A	/6
EV	/6
C	/4

Procedure:
1. Set up the apparatus as shown in the diagram
2. Hang the unknown mass and two slotted masses from the 3 threads and let the system come to equilibrium.
3. Displace it several times and notice the variation in the position of the point O.
4. When the system comes to rest in an approximately mean position, accurately mark the position of point O and the directions of the 3 forces on a piece of white paper fixed to the board.
5. Label the point of intersection of the 3 force vectors as O.
6. Repeat the above procedure for two more systems by changing the magnitude of coplanar forces acting on the body.
7. Record the value of forces A and B for each system.

Diagram:

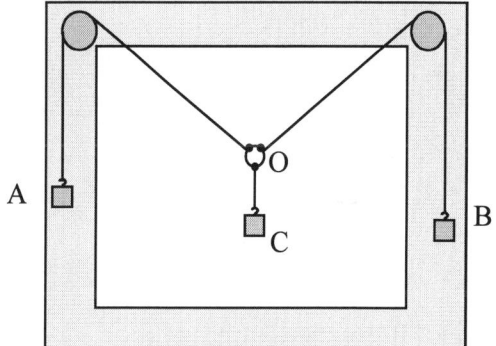

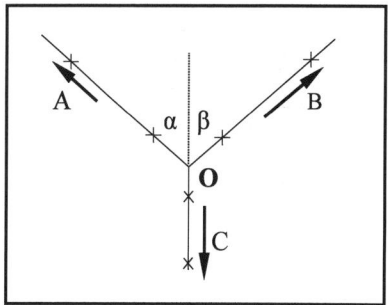

Analysis:
1. Attach your three A4 sheets which clearly show the original data obtained.
2. Using a suitable technique, determine the vector sum A+B and hence calculate the mass of the unknown object.
3. Compare this sum with the value of C obtained by measuring the unknown mass with a balance.
4. Include any uncertainties (both magnitude and direction).

INVESTIGATING THE FALL OF A COFFEE FILTER

Aim:
As a basket-type coffee filter falls, it tends to fall straight down and not flip over. This allows us to design a lab with several interesting variations on the independent and dependent variables. You were told recently to drop coffee filters and change several different variables as you dropped the filters. You also measured several different responding variables. Based on your brief exposure to the wonders of falling coffee filters, you will design a lab that could be performed by another student in a normal physics class.

Apparatus:
Basket-type coffee filters and any other equipment you think you might need.

IB Criteria Assessed

Criteria assessed	Level awarded
PE	/2
EX	/6
A	/6
EV	/6
C	/4

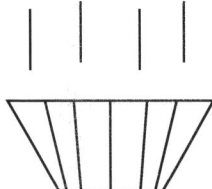

Practice Exploration:
Design a procedure to test a factor(s) that affects a falling coffee filter and that includes appropriate use of apparatus for the control, collection and analysis of data

This procedure should include the following sections.

- Defining the Problem and selecting variables:
- Controlling the Variables:
- Developing a procedure for collecting data:

Step by step instructions and diagrams are helpful to the reader and highly recommended. Also include a hypothesis and a sketch graph of what you think will happen.

Analysis:
- Record the raw data (both quantitative and qualitative) for the experiment in a suitable form. Include uncertainties due to the precision of the measuring apparatus.
- Process your quantitative raw data.
- Present the processed data in an appropriate way and include errors and uncertainties
- Draw conclusions based on your interpretation of the data

Evaluation:
- Evaluate your own plan, including any weaknesses and/or limitations. Include an evaluation of the apparatus used.
- In light of the weaknesses and limitations suggested above, suggest ways in which the procedure could be modified in order to improve it for the future.

INVESTIGATING PARABOLIC MOTION

Aim:
To calculate the initial horizontal take-off velocity of an object that is allowed to fall after being released from a ramp.

IB Criteria Assessed	
Criteria assessed	Level awarded
PE	/2
EX	/6
A	/6
EV	/6
C	/4

Procedure:
1. Set up the apparatus as shown below.
2. Place the wooden board in the vertical position so that it is touching the bottom of the ramp.
3. Release the ball bearing from height h so that it accelerates down the ramp and makes a mark on carbon paper attached to the wooden board.
4. Move the wooden board away from the ramp by a small distance (s_x) and release the ball once again.
5. Repeat for 6 to 8 distances of s_x.

Diagram:

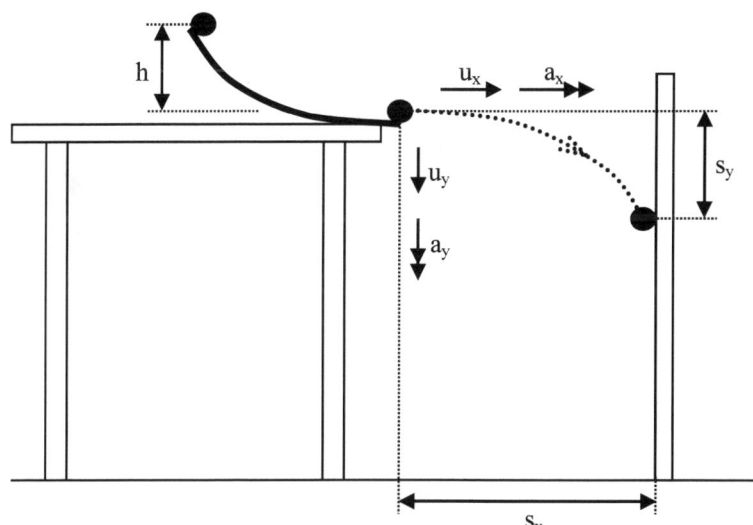

h = vertical height for the ball to be dropped.

u_x = initial horizontal velocity.

u_y = initial vertical velocity.

a_x = initial horizontal acceleration.

a_y = initial vertical acceleration.

s_x = horizontal distance.

s_y = vertical distance.

Analysis:
- Measure distances s_x and corresponding distances s_y and record these values in a suitable table. **Note: It is important to use fundamental units.**
- Include uncertainties due to the precision of the measuring apparatus.
- Process your quantitative raw data by drawing a suitable graph which will allow the initial horizontal take off velocity u_x to be calculated.
- Present the processed data in an appropriate way and include errors and uncertainties

Evaluation:
- Evaluate the data that you have collected and analyzed - compare your results by calculating the initial horizontal take off velocity (u_x) using the conservation of mechanical energy equations (mgh = ½mv²)
- Evaluate the procedure, including any modifications you had to make to overcome problems. Include an evaluation of the apparatus used.
- Suggest ways in which the procedure could be modified in order to improve it for the future.

Investigating the Flight of an Elastic Band

Aim:
A very important idea in scientific experiments is "controlling the variables". The aim is to keep all factors constant in the experiment except for the 2 you are testing. In this practical you will learn, for example, to examine the relationship between angle of launch and range while at the same time keeping other factors, such as amount of stretch of the band, constant. This is called a "fair test" in science.

Also, because of the randomness of the flight of the elastic band, you will also realize the importance of repeating your readings.

IB Criteria Assessed

Criteria assessed	Level awarded
PE	/2
EX	/6
A	/6
EV	/6
C	/4

Apparatus:
An elastic band plus any other equipment that you think you need.

Practice Exploration:
Design a procedure to investigate the flight of your elastic band that includes appropriate use of apparatus for the control, collection and analysis of data. This procedure should include the following sections.
- Defining the Problem and selecting variables:
- Controlling the Variables:
- Developing a procedure for collecting data:

Include a quantitative hypothesis for your investigation.

Analysis:
- Record the raw data (both quantitative and qualitative) for the experiment in a suitable form. Include uncertainties due to the precision of the measuring apparatus.
- Process your quantitative raw data.
- Present the processed data in an appropriate way and include errors and uncertainties
- Draw conclusions based on your interpretation of the data

Evaluation:
- Evaluate your own plan, including any weaknesses and/or limitations. Include an evaluation of the apparatus used.
- In light of the weaknesses and limitations suggested above, suggest ways in which the procedure could be modified in order to improve it for the future.

INVESTIGATING THE SIMPLE PENDULUM

Aim:
This investigation is designed to give you an introduction to IB exploration, analysis and evaluation experiments.

You should design an experiment in order to test different variables that may contribute to the time period of a pendulum and try to determine a mathematical relationship between these variables.

Apparatus:
Whatever you feel is suitable for this experiment.

IB Criteria Assessed

Criteria assessed	Level awarded
PE	/2
EX	/6
A	/6
EV	/6
C	/4

Practice Exploration:
Design a procedure to test a factor(s) that affects the time period of a Simple Pendulum that includes appropriate use of apparatus for the control, collection and analysis of data. This procedure should include the following sections.
- Defining the Problem and selecting variables:
- Controlling the Variables:
- Developing a procedure for collecting data:

Include a quantitative hypothesis for your investigation.

Analysis:
- Record the raw data (both quantitative and qualitative) for the experiment in a suitable form. Include uncertainties due to the precision of the measuring apparatus.
- Process your quantitative raw data.
- Present the processed data in an appropriate way and include errors and uncertainties
- Draw conclusions based on your interpretation of the data

Evaluation:
- Evaluate your own plan, including any weaknesses and/or limitations. Include an evaluation of the apparatus used.
- In light of the weaknesses and limitations suggested above, suggest ways in which the procedure could be modified in order to improve it for the future.

INVESTIGATING THE STOPPING DISTANCE OF A BICYCLE

Aim:
Safety on the road is an important part of everyday life. Cars are capable of faster and faster speeds and these higher speeds bring the need for better brakes, better tires, and better roads.

In an emergency situation, perhaps a person stepping off the curb and into the road, a driver has to be able to stop the vehicle before reaching the person. There are numerous factors which might affect this stopping distance.

Your task is to identify these factors, choose ONE to test and find a relationship between your chosen variable and the stopping distance.

IB Criteria Assessed

Criteria assessed	Level awarded
PE	/2
EX	/6
A	/6
EV	/6
C	/4

Apparatus:
One bicycle plus anything else you need to perform the experiment to your plan.

Practice Exploration:
Design a procedure to test a factor that affects the stopping distance of a bicycle that includes appropriate use of apparatus for the control, collection and analysis of data.

This procedure should include the following sections.

- Defining the Problem and selecting variables:
- Controlling the Variables:
- Developing a procedure for collecting data:

Include a quantitative hypothesis for your investigation.

Analysis:
- Record the raw data (both quantitative and qualitative) for the experiment in a suitable form. Include uncertainties due to the precision of the measuring apparatus.
- Process your quantitative raw data.
- Present the processed data in an appropriate way and include errors and uncertainties
- Draw conclusions based on your interpretation of the data. Include a comparison with published data relating the stopping distance of a vehicle and the variable that you tested.

Evaluation:
- Evaluate your own plan, including any weaknesses and/or limitations. Include an evaluation of the apparatus used.
- In light of the weaknesses and limitations suggested above, suggest ways in which the procedure could be modified in order to improve it for the future.

INVESTIGATING THE TORSIONAL PENDULUM

Aim:
The time period of a pendulum was first investigated by Christian Huygens in the 17th Century (after apparently observing a swinging chandelier in church). From Huygen's work the first accurate clocks were invented and refined

You may be very familiar with traditional swinging pendulums and the variables that do and do not affect their time periods. In this design lab, you will learn about the workings of a torsional (twisting) pendulum.

IB Criteria Assessed

Criteria assessed	Level awarded
PE	/2
EX	/6
A	/6
EV	/6
C	/4

Diagram:

Apparatus:

- A torsional pendulum made from a 30cm acrylic ruler and elastic band (shown opposite)
- Anything else you might find in a normal physics classroom.

Practice Exploration:
Design a procedure to test how a certain variable (of your choice) may affect the rate of an oscillating torsional pendulum that includes appropriate use of apparatus for the control, collection and analysis of data. As always, this should design lab should include:

- Defining the Problem and selecting variables:
- Controlling the Variables:
- Developing a procedure for collecting data:

Step by step instructions and diagrams are helpful to the reader and highly recommended.

Also include a quantitative hypothesis for your investigation together with a sketch graph of what you think will happen.

Analysis:
- Record the raw data (both quantitative and qualitative) for the experiment in a suitable form. Include uncertainties due to the precision of the measuring apparatus.
- Process your quantitative raw data.
- Present the processed data in an appropriate way and include errors and uncertainties
- Draw conclusions based on your interpretation of the data. Include a comparison of your results with any published data regarding torsional pendulums. Compare your results with your original hypothesis.

Evaluation:
- Evaluate your own plan, including any weaknesses and/or limitations. Include an evaluation of the apparatus used.
- In light of the weaknesses and limitations suggested above, suggest ways in which the procedure could be modified in order to improve it for the future.

Original Lab Sheet by Ringo Dingrando

INVESTIGATING NEWTON'S 2ND LAW OF MOTION USING A TICKER TIMER

Aim:
To use the ticker timer and ticker tape to verify Newton's 2nd Law of Motion (Law of Inertia)

IB Criteria Assessed

Criteria assessed	Level awarded
PE	/2
EX	/6
A	/6
EV	/6
C	/4

Apparatus:
1 runway, 9 x 10g masses plus holder, string, scissors, dynamics trolley, tickertape and timer, lab power supply, 2 x long wires, wedge, pulley, top pan balance.

Procedure:
1. Set up the apparatus as shown below. Connect the ticker timer to a 12V a.c. supply.
2. The slope of the runway should be adjusted so that the trolley runs at a constant speed down the slope when pushed (as judged by the eye). In this condition there is no resultant force acting on the trolley along the slope. The runway has now been "friction compensated"
3. Measure the mass of the trolley
4. Cut a length of tickertape (about 1m), attach it to the trolley.
5. Hang a mass holder, as shown. This has a mass of 10g and therefore pulls the trolley with a force of 0.098N down the slope. With the ticker timer running, allow the trolley to be pulled down the slope. CATCH THE TROLLEY BEFORE IT FALLS OFF THE END OF THE RUNWAY.
6. Repeat the above using different masses to create different pulling forces.

Diagram:

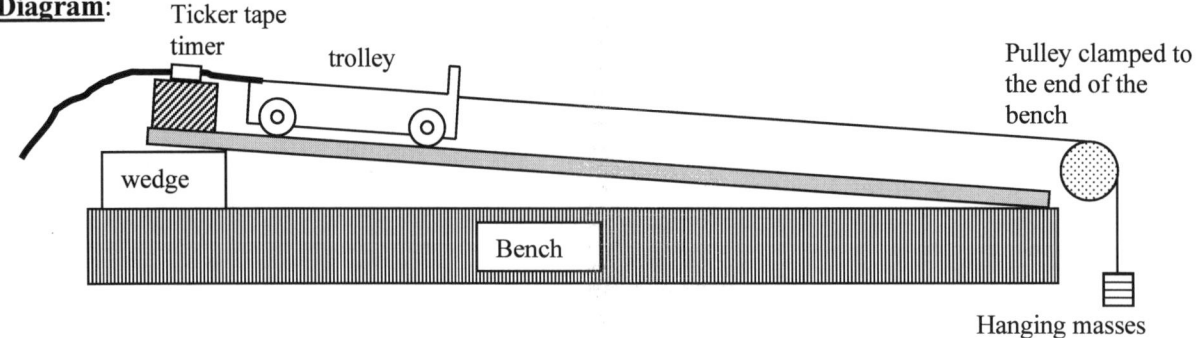

Theory:
Distance $s = ut + \frac{1}{2}at^2$ $\therefore a = 2s/t^2$ (initial velocity $u = 0$)
Ticker timer produces 50 "ticks" per second $\therefore$ time between ticks = 0.02s (50 ticks = 1s)
Measuring a length of tape which is x ticks long will allow you to calculate the actual acceleration.

Analysis:
- Collect data from the ticker tape and record the forces which were used for each piece of tape. Present these data clearly.
- Calculate the expected acceleration in each case using Newton's 2nd Law.
- Use the ticker timers to determine the actual accelerations.
- Give a conclusion and explanation of your results, compare the two values for acceleration in each case and make appropriate comments.

Evaluation:
- Evaluate the above procedure and apparatus used, including limitations, weaknesses or errors.
- Suggest ways of improving the investigation for the future.

INVESTIGATING HOOKE'S LAW

Aim:
Robert Hooke was one of the first to notice a relationship between the force applied to an elastic object and its extension. This lab is designed to test and verify Hooke's Law, which states that "the extension of an elastic material is directly proportional to the applied force so long as the elastic limit is not exceeded".

Apparatus:
Clamp and stand, spring, meter rule, mass holder, slotted masses

IB Criteria Assessed

Criteria assessed	Level awarded
PE	/2
EX	/6
A	/6
EV	/6
C	/4

Procedure:
1. Set up the apparatus as shown in the diagram.
2. Apply various forces to the spring
3. Record its extension
4. (Remember that the extension = final length − original length).

Diagram:

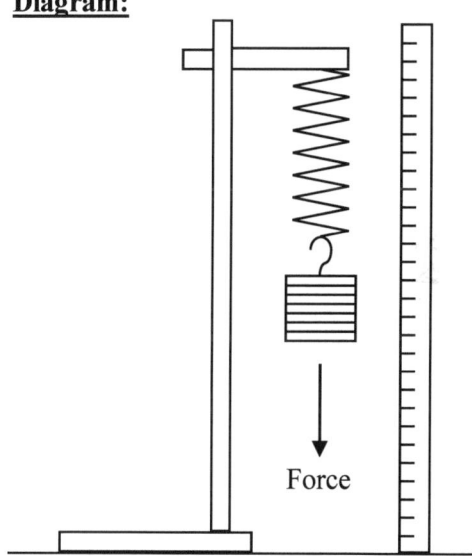

Analysis:
- Record pairs of data for force and extension in a suitable table.
- Your results table and the presentation of data should include any uncertainties associated with the apparatus that you have used.
- Use your results to plot a suitable graph which will allow you to accurately calculate the spring constant, k, for your spring.
- SL & HL, include uncertainty bars for each data point.
- HL, calculate the uncertainty in your final value for the spring constant by drawing maximum and minimum slope on your graph.

Evaluation:
- Your evaluation should include a final value for the spring constant of your spring together with an estimate of the uncertainty.
- Does your graph verify Hooke's Law?
- Evaluate the procedure and result including limitations, weaknesses or errors.
- Suggest possible causes of errors and modifications that could be introduced to improve the investigation.

INVESTIGATING SPRINGS

Aim:
To investigate the interaction of various spring systems. (Examples of different spring systems are shown in the diagrams below).

Diagrams (Explanation of series and parallel spring systems):

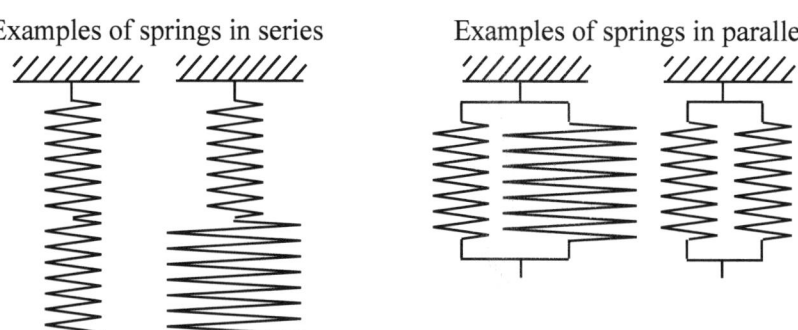

IB Criteria Assessed	
Criteria assessed	Level awarded
PE	/2
EX	/6
A	/6
EV	/6
C	/4

Apparatus:
Various springs of different spring constants. Any other apparatus that you can think of to complete the investigation.

Practice Exploration:
Design a procedure that includes appropriate use of apparatus for the control, collection and analysis of data. This procedure should include the following sections. Include a quantitative hypothesis for your investigation.

Analysis:
- Show the results for the experiment in a suitable table. Include uncertainties.
- Present all of your results in the form of suitable graphs
- Suggest relationships to help explain the systems analyzed.

Evaluation:
- Evaluate the data that you have collected and analyzed - compare your results to literature values.
- Evaluate your own plan, including any modifications you had to make to overcome problems. Include an evaluation of the apparatus used.
- Suggest ways in which the procedure could be modified in order to improve it for the future.

INVESTIGATING WORK DONE AND ENERGY TRANSFERRED ON AN INCLINED PLANE

Aim:
1. To find a relationship between force applied and distance moved up an inclined plane
2. To use this relationship to prove that **Work done = Energy transferred**
3. To use raw data as an exercise in data manipulation in order to calculate the height, h

IB Criteria Assessed

Criteria assessed	Level awarded
PE	/2
EX	/6
A	/6
EV	/6
C	/4

Diagram:

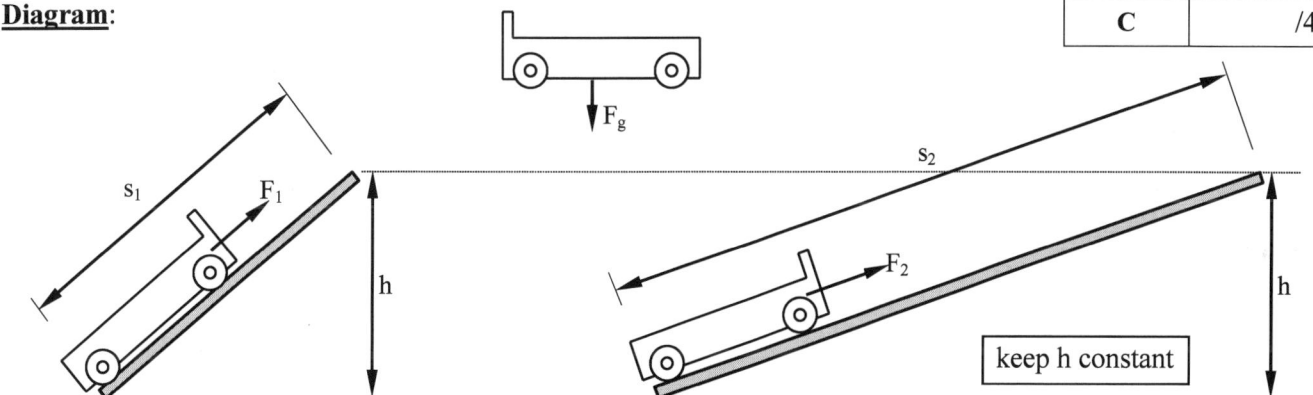

Procedure:
1. Set up the apparatus as shown in the diagram below (any value for h will do).
2. Measure the height of the ramp, h and record it.
3. Measure the weight of the dynamics trolley, F_g and record it.
4. Using a force meter, pull the trolley up the ramp (maintaining a constant velocity), recording the length of the ramp, s, and the force required, F.
5. Change the angle of the ramp and hence the distance, s and force, F. **KEEP h CONSTANT.**
6. Repeat steps 2 to 5 above until you are satisfied with the quantity of collected data.

Analysis:
- Collect and record pairs of results for s and F including units and uncertainties.
- Present these data clearly.
- Process your raw data in a way which will allow you to accurately calculate the value of h (the height of the ramp). Refer to the "Aim" of the investigation.
- Include any errors or uncertainties in your processed data.
- Give a conclusion and explanation of your results. Compare your calculated value of h to the actual value (measured from the apparatus)

Evaluation:
- Evaluate the above procedure and apparatus used, including limitations, weaknesses or errors.
- Suggest ways of improving the investigation.

Original Lab Sheet by Mike Dickinson

INVESTIGATING THE BALLISTIC PENDULUM

Aim:
To investigate the laws of conservation of momentum and energy.

Theory:
When a projectile is fired at the pendulum and captured by it, the combination of pendulum and projectile will swing upwards by an amount that is indirectly related to the original velocity of the bullet. According to the law of conservation of energy, the increase in height, Δh, of the pendulum block and projectile is such that the increase in potential energy is equal to the kinetic energy just after the collision.

IB Criteria Assessed

Criteria assessed	Level awarded
PE	/2
EX	/6
A	/6
EV	/6
C	/4

Diagram:

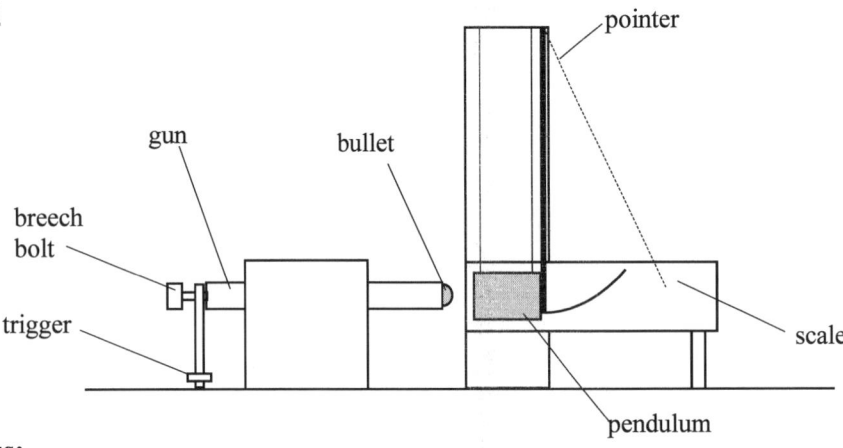

Apparatus:
Ballistic pendulum (model P62401)

Procedure:
1. Measure the mass of the pendulum and the mass of the bullet using an electronic balance.
2. Measure the length of the pendulum with a ruler.
3. Pull the breech bolt until it is latched by the trigger at the 1st position, load the gun with the bullet and fire the gun so that the bullet gets lodged inside the pendulum
4. Record the position of the pointer in degrees.
5. Repeat step 3, 5 times using the same notch on the breech bolt
6. Repeat steps 3, 4 and 5, using the 2nd and 3rd latch positions

Analysis:
- In a suitable table record all of the data collected.
- Include actual and percentage uncertainties due to the measuring apparatus used.
- Analyze the raw data in an appropriate way that will accurately determine the velocities of the bullet at the 3 breech bolt positions.

Evaluation:
- Evaluate the results of the experiment. Suggest ways of checking the reliability of your analysis.
- Evaluate the procedure, including any modifications you had to make to overcome problems. Include an evaluation of the apparatus used.
- Suggest ways in which the procedure could be modified in order to improve it for the future.

INVESTIGATING THE POWER AND TEMPERATURE OF THE SUN

Aim:
In many situations it is impossible to measure something directly. This experiment is a clever example of how to use physics theory, and some mathematics, to measure indirectly, what is impossible to do directly - namely to find the power and temperature of the sun.

IB Criteria Assessed

Criteria assessed	Level awarded
PE	/2
EX	/6
A	/6
EV	/6
C	/4

Procedure:
1. Measure the mass of the test tube and the mass of the water.
2. Take the temperature of the water.
3. Focus the sun on to the water and stir occasionally.
4. Hold the lens and test tube steady until the water has increased in temperature a reasonable amount (do not allow the water to boil).

Diagram:

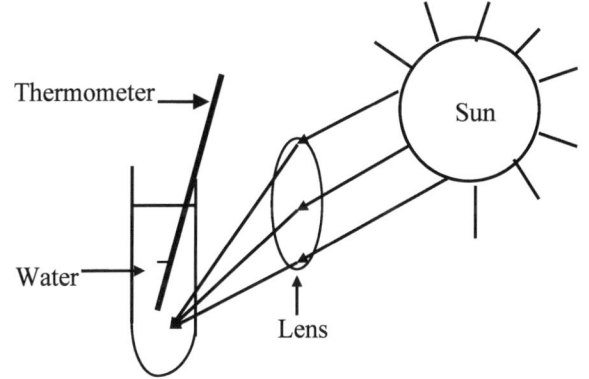

Apparatus:
Magnifying glass
Thermometer
Test tube
Stopclock.

Theory:
The energy that passes through the lens heats the water and the test tube. This energy can be calculated using the equation for specific heat capacity. This is the energy passing through area A m^2 of the lens. Knowing the radius of orbit of the Earth around the Sun, you can calculate the total energy emitted by the Sun. The surface temperature of the Sun can be calculated by using the "Stefan - Boltzmann" equation.

$$\text{Energy / second / m}^2 \text{ from the Sun's surface} = 5.7 \times 10^{-8} \times (\text{Kelvin temperature})^4$$

Analysis:
- Collect data that will allow you to calculate the surface temperature and power of the sun.
- Include units and uncertainties.
- Present these data clearly.
- Using the above procedure and theory, calculate the surface temperature and power of the sun.

Evaluation:
- The information you need can all be found in your data book (the true value for the power of the Sun is 3.9×10^{26} W). Compare your calculated value to this literature value
- There are many faults in this experiment. Try to think of some of them. If you have time then repeat the experiment but with an improved method.
- Evaluate the procedure including any modifications you had to make to overcome problems. Include an evaluation of the apparatus used.
- Suggest ways in which the procedure could be modified in order to improve it for the future.

Physics Practical Scheme of Work – For use with the IB Diploma Programme – First Assessment 2016

INVESTIGATING RATE OF COOLING

Introduction:
Is the rate of cooling in still air $\propto$ (excess temperature)$^{5/4}$?

Aim:
You have a law expressed mathematically and there are 2 useful skills here for you to learn:
1. You will need to use your knowledge of logs to manipulate the data so that a linear relationship can be established.
2. Having collected your data you need to plot the appropriate graph in order to test the 5/4 relationship.

IB Criteria Assessed

Criteria assessed	Level awarded
PE	/2
EX	/6
A	/6
EV	/6
C	/4

Apparatus:
Beaker of water, heater, stopclock, thermometer.

Procedure:
1. Heat the water to boiling point
2. Measure and record the temperature of the water
3. Start the stopclock and remove the water from the heat
4. Record the temperature of the water each minute as it cools

Hint:
Notice that the law says "<u>rate</u> of cooling". This means "°C per minute". This is important when you decide what graph you are going to plot to test the law of cooling. Remember also that before you decide if the law of cooling is true or not, you must decide if your experiment was accurate enough to justify your conclusions.

Analysis:
- Collect pairs of date for temperature and time as the water cools
- Show the results for the experiment in a suitable table. Include uncertainties.
- Analyze your raw date in such a way that will allow a conclusion to be reached about the relationship between rate of cooling and excess temperature.

Evaluation:
- Evaluate the data that you have collected and analyzed - compare your results to the value given.
- Evaluate your own plan, including any modifications you had to make to overcome problems. Include an evaluation the apparatus used.
- Suggest ways in which the procedure could be modified in order to improve it for the future.

DETERMINING THE TEMPERATURE OF A WIRE BY EXPANSIVITY

Aim:
To measure the temperature of a hot wire is very difficult to do directly - you can't simply touch a thermometer against the wire. This experiment is interesting because you use one property (thermal expansion) in order to measure another (temperature). This indirect technique is used a lot in science - for example the temperature of a star can be measured by examining its color.

Also in the real world, engineers have to know the effect of current through wires in order to allow for sag in high voltage overhead cables

IB Criteria Assessed	
Criteria assessed	Level awarded
PE	/2
EX	/6
A	/6
EV	/6
C	/4

Apparatus:
1 m of copper wire, 2 clamp stands, four 100 g masses, variable power supply, ammeter, sheet of plane paper, leads, crocodile clips.

Diagram:

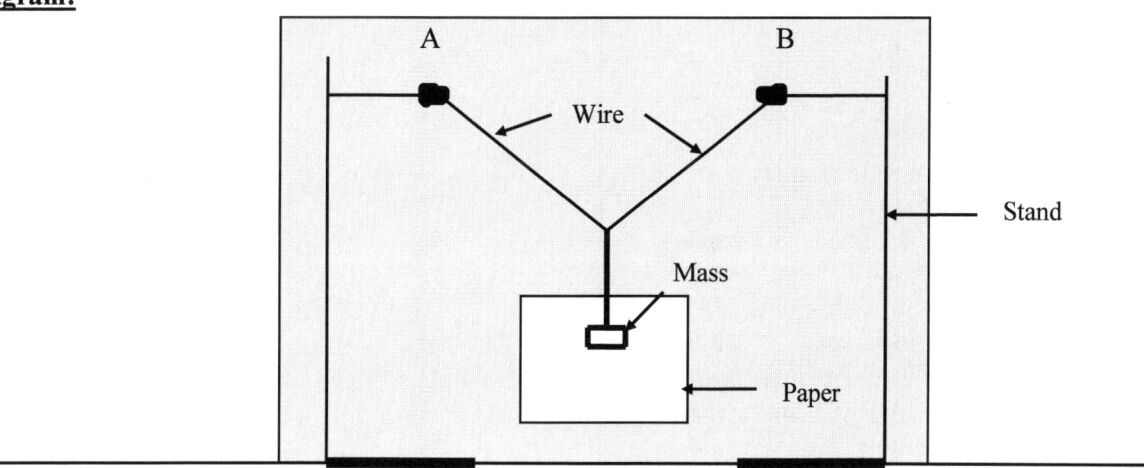

Procedure:
1. Clamp the wire firmly at A and B using the 100 g masses as grips.
2. Hang the mass from the wire and mark its position on a piece of blank paper.
3. Connect the power supply with the ammeter in series, to A and B.
4. Set the current to 1A and mark the new position of the hanging mass.
5. Repeat this in steps of 1 A, up to 5A.
6. Use the lines on the paper to measure the increase in length of the wire for each reading and, knowing the thermal expansivity of the wire's material (look this up in a data book), calculate the average temperature of the wire for each of the 5 different currents.

Analysis:
- Collect and record pairs of data for current and average temperature.
- Present these data in a results table – include the uncertainties in these two variables
- Plot a suitable graph to see the relationship between current and average temperature. Include uncertainties.
- Look up the melting point of copper and use your graph to estimate what current will melt the wire.
- Include an estimate in the uncertainty of your calculated value.

INVESTIGATING PING-PONG BALLS

Aim:
This is a real - life problem. At Wimbledon tennis tournament, the balls are kept in a fridge before the players use them. This is because the ball's performance changes enough with temperature to affect their bounciness and this could have a serious effect on the result of the game - enough maybe for example to just put the ball over the line.

You will learn that investigating a real - life situation can become very complicated. "How hard do table tennis players hit the ball?" It is often very difficult to simulate in the laboratory what actually happens in the real world.

IB Criteria Assessed

Criteria assessed	Level awarded
PE	/2
EX	/6
A	/6
EV	/6
C	/4

Apparatus:
3 ping - pong balls plus any other equipment you think will be necessary.

Practice Exploration:
This is an open ended experiment so you can do anything you want, but remember that you will have to show to the "International Ping - Pong Federation" that you have thoroughly investigated all aspects of the problem and that you are certain of your conclusions. Design a procedure that includes appropriate use of apparatus for the control, collection and analysis of data. This procedure should include the following sections:

- Defining the Problem and selecting variables:
- Controlling the Variables:
- Developing a procedure for collecting data:

Include a quantitative hypothesis for your investigation.

Analysis:
- Show the results for the experiment in a suitable table. Include uncertainties.
- Analyze your raw data in such a way that will allow a conclusion to be reached about the relationship between your selected variable and the bounciness of the Ping-Pong balls.

Evaluation:
- Evaluate the data that you have collected and analyzed - compare your results to literature values if possible.
- Evaluate your own plan, including any modifications you had to make to overcome problems. Include an evaluation of the apparatus used.
- Suggest ways in which the procedure could be modified in order to improve it for the future.

INVESTIGATING THE PRESSURE / VOLUME RELATIONSHIP FOR A BALLOON

Aim:
This experiment is different because you have to try and formulate some sort of relationship between the pressure and the volume of a balloon but you have no idea what it might be so you will have to confront a completely original problem for which there is no answer in a text book. Good luck!

IB Criteria Assessed

Criteria assessed	Level awarded
PE	/2
EX	/6
A	/6
EV	/6
C	/4

Procedure:
1. Using first the round balloon, partially inflate it and connect it to the manometer.
2. The difference in height of water between the 2 sides gives the pressure (use "p = ρ x g x h" to convert to Pascals).
3. How to measure the volume of the balloon is your decision.
4. You need to use a bit of ingenuity here.
5. Inflate the balloon a bit more and once again measure pressure and volume.
6. Obtain a total of 5 or 6 readings.
7. Now do the whole experiment again but this time using the "sausage" balloon.

Apparatus:
Manometer, 1 "round" balloon, 1 "sausage" balloon, plus whatever else you think you need.

Diagram:

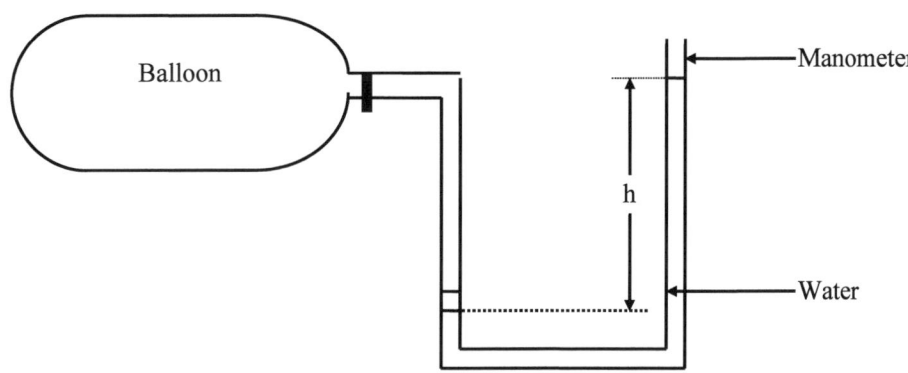

Analysis:
- Collect whatever data you feel is necessary in order to complete the data processing parts below.
- Plot a suitable graph which will allow you to find the relation between pressure and volume - e.g. "p" against "$v^{1/2}$".
- Remember that you are looking for a straight line graph.
- Do both balloons give the same result?

INVESTIGATING MALUS' LAW

Aim:
As unpolarized light passes through a polarizing filter, the Polaroid removes all the light whose axes of polarization are not in-line with the plane of polarization of the Polaroid. This leaves only light with electric field oscillations in one orientation. The light is said to be "plane polarized". The intensity of the polarized light is half the original source intensity, $I = \frac{1}{2}I_0$

If a second Polaroid, known as an analyzer, is placed in front of the first Polaroid, the intensity of the light can be controlled according to Malus' Law. The intensity is seen to be a function of the angles of polarization between the first and second Polaroids.

IB Criteria Assessed

Criteria assessed	Level awarded
PE	/2
EX	/6
A	/6
EV	/6
C	/4

1. To verify the use of a Polaroid to halve the intensity of the original light source.
2. To verify Malus' Law using crossed Polaroids.

Theory:
The relationship between the intensity of the light leaving the second (crossed) Polaroid is given by Malus' Law.

$$I = I_0 \cos^2 \theta$$

where I = intensity of the transmitted beam
I_0 = intensity of the incident (plane polarized) beam
θ = angle between the plane of polarization of the incident beam and the plane of Polaroid's axis

Diagram:

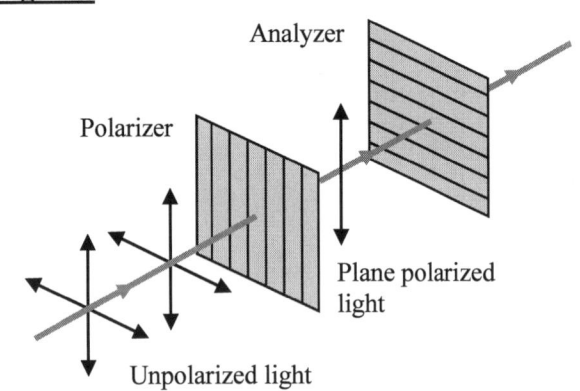

Procedure:
1. Set up the apparatus as shown in the diagram.
2. Start with the 2 Polaroids giving the same plane of polarization (this can be achieved by lining the 0° mark on both Polaroids at the 12 o'clock position in their holders.
3. Adjust the angle between the Polaroids in 5° increments until their planes of polarization are perpendicular (as shown in the diagram)
4. Record pairs of data for angle and light intensity, until an angle is reached where the intensity of the light falls to a zero (or a minimum).
5. Continue past this minimum – the intensity of the light should start to increase once again.
6. Record the uncertainties in the collected data.

Analysis:
- Collect appropriate data and present this in a way which will allow for easy interpretation.
- Include the uncertainties of measurement in your recorded data.
- Plot a suitable graph from which you can analyses the intensity and degree of polarization of the polarized light in order to verify Malus' Law.
- Include the uncertainty in your calculated answer.
- Make valid conclusions related to intensity of plane polarized light and Malus' Law

Evaluation:
- Explain the results in terms of degree of polarization and angles of the crossed Polaroids
- Evaluate the procedure, including any modifications you had to make to overcome problems
- Include an evaluation of the apparatus used.
- Suggest ways in which the procedure could be modified in order to improve future investigations.

INVESTIGATING BREWSTER'S LAW

Aim:
As light reflects from the surface of a non-metallic substance, it becomes partially plane-polarized and the degree of polarization depends upon the viewing angle of the reflected light. At a certain angle, the light will be completely plane polarized and this occurs at "Brewster's Angle" (named after the Scottish physicist, Sir David Brewster). It is possible to calculate the refractive index of the reflecting medium using Brewster's angle. The aims of this experiment are:

1. To verify Brewster's Law for (partially) plane-polarized reflected light.
2. To use the data collected to find the refractive index of a piece of Acrylic.

IB Criteria Assessed

Criteria assessed	Level awarded
PE	/2
EX	/6
A	/6
EV	/6
C	/4

Theory:
The relationship between refractive index and Brewster's angle can be given by
$n = \tan \theta_B$

> SAFETY FIRST – DO NOT LOOK DIRECTLY AT THE LASER LIGHT

Diagram:

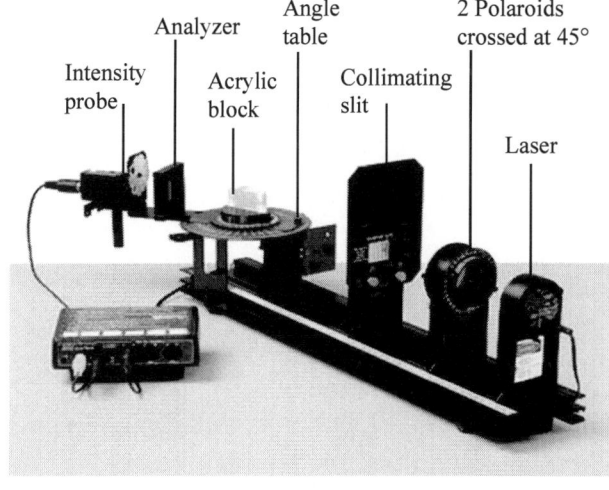

Procedure:
1. Set up the apparatus as shown in the photograph. (The laser light is already polarized, so the 2 Polaroids crossed at 45° unpolarizes this laser beam).
2. Direct the laser beam onto the surface of the acrylic plastic block.
3. Set the angle of the analyzer so that it is oriented horizontally, to eliminate the unpolarized reflected light.
4. Change the angle at which the laser beam is incident on the surface of the plastic and make necessary adjustments so that the reflected beam enters the intensity probe.
5. Record pairs of data for angle and intensity, until an angle is reached where the intensity of the reflected light falls to a minimum.
6. Continue past this minimum – the intensity of the reflected light should start to increase once again.
7. Record the uncertainties in the collected data.

Analysis:
- Collect appropriate data and present this in an appropriate way, allowing for easy interpretation.
- Plot a suitable graph from which you can analyze the intensity and degree of polarization of the reflected light for various angles of reflection.
- Use the graph to determine Brewster's angle and the refractive index for the Acrylic. Include the uncertainty in your calculated answers.

Evaluation:
- Quote your values found for Brewster's angle and refractive index for Acrylic and compare these to literature values.
- Explain the results in terms of degree of plane-polarization from a non-metallic reflector's surface.
- Evaluate the procedure, including any modifications you had to make to overcome problems. Include an evaluation of the apparatus used.
- Suggest ways in which the procedure could be modified in order to improve it for the future.

Original Lab Sheet by Ringo Dingrando and Mike Dickinson

Determining the Wavelength of Light using a Diffraction Grating

Aim:
To determine the wavelengths of red, green, and blue light by experimental method using a diffraction grating.

Procedure:
1. Place the lamp, lens and screen in the positions shown in the diagram.
2. Adjust the lens distance to form a focused image on the screen.
3. Place the diffraction grating in the position shown.
4. Adjust the diffraction grating until clear fringes are formed on the screen.
5. Measure the length l and corresponding distance x and record them in a suitable table. (When using white light, three measurements can be made for x – from the central bright fringe to the red, green and blue parts of the first fringe).
6. Repeat for various distances of l and x.

IB Criteria Assessed

Criteria assessed	Level awarded
PE	/2
EX	/6
A	/6
EV	/6
C	/4

Theory:

$$\text{Wavelength } n \lambda = d \sin \theta$$

where d = diffraction grating slit width
 d = 1/N
 N = number of lines per mm

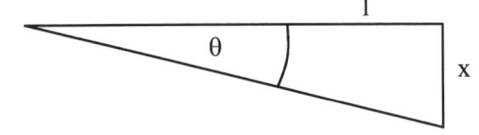

But for small angles of θ, $\tan \theta \cong \sin \theta$

$$\therefore \text{Wavelength } n \lambda = \frac{d x}{l}$$

Diagram:

Analysis:
- Collect of pairs of results for l and x for the colors red, green and blue.
- Present these (together with uncertainties) in an appropriate table.
- Plot a suitable graph which, combined with the theory above, will enable you to calculate the wavelengths of the three colors of light.

Evaluation:
- Evaluate your results.
- Compare your values for the wavelength of the 3 colors of light to those quoted in the text book.
- Comment on any possible sources of error, especially regarding the small angle approximation used in the theory above.
- Identify any weaknesses and suggest ways of improving the investigation.

INVESTIGATING MELDE'S EXPERIMENT

Aim:
In this experiment you will collect raw data on frequency and tension in a vibrating string and your task is to try and find some simple mathematical relationship between them. To collect data and, from it, produce an equation which we can use to make predictions about the behavior of a system, is a very important process in science.

IB Criteria Assessed

Criteria assessed	Level awarded
PE	/2
EX	/6
A	/6
EV	/6
C	/4

Apparatus:
Signal generator, vibrator, leads, 3 m of cotton, eight 50 g masses, mass hanger, pulley, ramp.

Diagram:

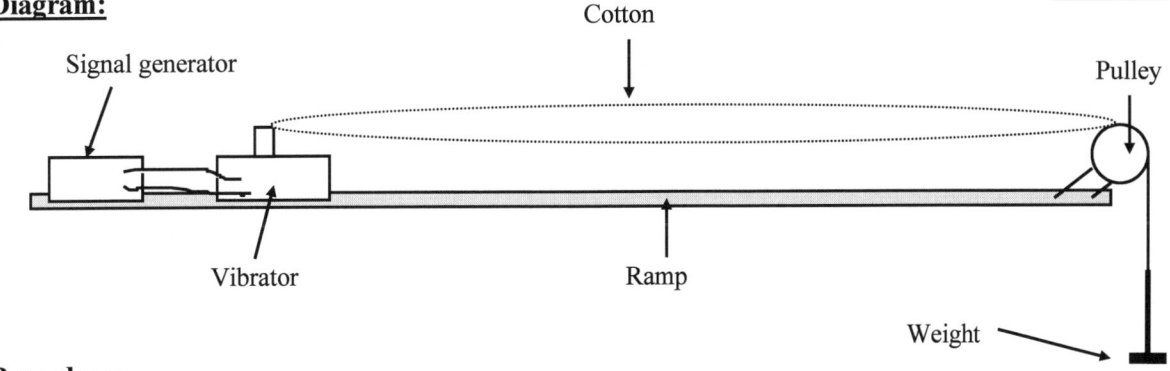

Procedure:
Put a 50 g mass on the string and adjust the signal generator until you get a large amplitude wave (as shown in the diagram). Note the frequency of the signal generator. Repeat this several times in steps of 50 g.

Theory:
There is clearly some sort of relationship between "frequency" and "tension", but what is it? It might be one of the following:

- frequency $\propto$ Tension in string
- frequency $\propto$ (Tension in string)^2
- frequency^2 $\propto$ Tension in string

Analysis:
- Collect of pairs of results for f and T for when the string gives a large single amplitude
- Present these data (together with uncertainties) in an appropriate table.
- Plot a suitable graph which will enable you to determine the relationship between the fundamental frequency and the tension in the string

Evaluation:
- Evaluate the data that you have collected and analyzed. What does the textbook say about the relationship between the fundamental frequency and the tension in the string? Compare your conclusion to this.
- Evaluate the procedure, including any modifications you had to make to overcome problems. Include an evaluation the apparatus used.
- Suggest ways in which the procedure could be modified in order to improve it for the future.

Original Lab Sheet by Brian Seve – Modified to the current IB syllabus requirements by Mike Dickinson

INVESTIGATING THE POWER OF AN ELECTRIC HEATER

Aim:
1. To obtain a relationship between the power dissipated by an electric heater, its resistance and the current flowing.
2. To use appropriate mathematics to decide if your results prove or disprove the equation.

Apparatus:
50 cm of resistance wire, variable D.C. power supply, ammeter, voltmeter, thermometer, stopwatch, leads, 250 cm^3 beaker.

IB Criteria Assessed

Criteria assessed	Level awarded
PE	/2
EX	/6
A	/6
EV	/6
C	/4

Diagram:

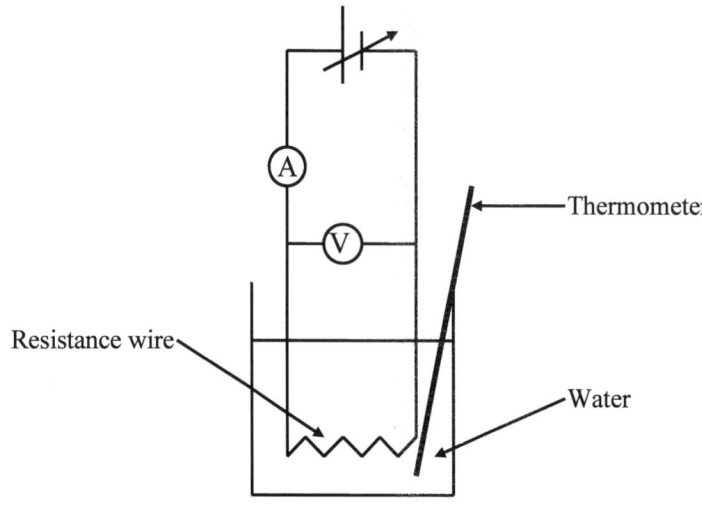

Procedure:
To produce the resistance coil, wrap the wire round a pencil. Set up the circuit. Take the temperature of the water. Set the current to 1 A, stir the water occasionally, and after 5 minutes take the new temperature Repeat the process at various currents.

Analysis:
- Collect and record pairs of results for Voltage and Current including units and uncertainties.
- Present these data clearly.
- Process your raw data in a way which will allow you to accurately calculate the power received by the water.
- Take into account any errors or uncertainties in your processed data.

INVESTIGATING RESISTORS IN SERIES AND PARALLEL

Aim:
To investigate the interaction of various sizes of resistors. (Examples of different resistor circuits are shown in the diagrams below).

IB Criteria Assessed	
Criteria assessed	Level awarded
PE	/2
EX	/6
A	/6
EV	/6
C	/4

Diagrams (Explanation of series and parallel resistor circuits):

A circuit with resistors in series

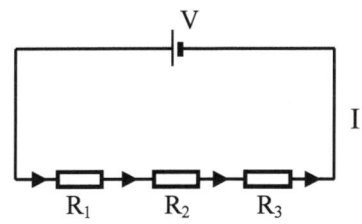

A circuit with resistors in parallel

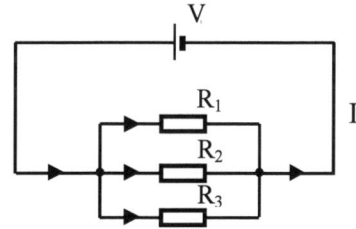

Apparatus:
Various resistors of different values. Any other apparatus that you can think of to complete the investigation.

Practice Exploration:
Design a procedure that includes appropriate use of apparatus for the control, collection and analysis of data. This procedure should include the following sections. Include a quantitative hypothesis for your investigation.

Analysis:
- Show the results for the experiment in a suitable table. Include uncertainties.
- Present all of your results in the form of suitable graphs
- Suggest relationships to help explain the systems analyzed.

Evaluation:
- Evaluate the data that you have collected and analyzed - compare your results to literature values.
- Evaluate your own procedure, including any modifications you had to make to overcome problems. Include an evaluation of the apparatus used.
- Suggest ways in which the procedure could be modified in order to improve it for the future.

Original Lab Sheet by Mike Dickinson

Verifying the Equation "F = B I L" Using a Current Balance

Aim:
The 2 benefits of this experiment are:
- it is important to know how to carry out a controlled scientific experiment that will step by step isolate and test each of the variables in an equation - in this case, "B", "I" and "L". ("controlling the variables")
- this experiment needs a lot of careful manipulation of the equipment to make it work perfectly. This precision and patience is vital in scientific experiments.

IB Criteria Assessed

Criteria assessed	Level awarded
PE	/2
EX	/6
A	/6
EV	/6
C	/4

Apparatus:
2 Magnadur magnets and holders, 1 m of thick copper wire, variable d.c. power supply, 2 metal pivots, crocodile clips, 50 cm of ticker tape.

Diagram:

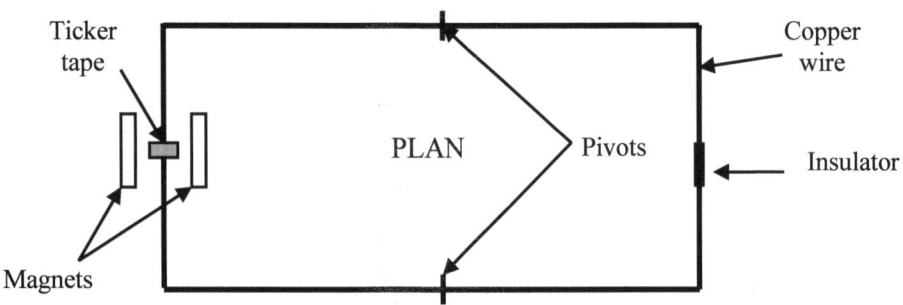

Procedure:
The copper frame is adjusted until it balances horizontally. Crocodile leads are connected to the 2 pivots and current passes through the left hand side of the copper frame (not the right hand side because the insulator blocks the current). The current direction is chosen so that the frame is pushed upwards by the electromagnetic force. A small piece of ticker tape is added so that the frame once again balances horizontally.

Theory:
The upward force (F) on the copper wire is given by the equation, F = B x I x L.

Analysis:
- Collect and record pairs of results for Force and Current including units and uncertainties.
- Present these data clearly.
- Process your raw data in a way which will allow you to accurately calculate the value of the magnetic field strength, B.
- Take into account any errors or uncertainties in your processed data.
- Give a conclusion and explanation of your results; compare to literature values if possible.

Evaluation:
- Evaluate the above procedure and apparatus used, including limitations, weaknesses or errors.
- Identify any weaknesses and suggest ways of improving the investigation.

INVESTIGATING ELECTROMAGNETS

Aim:
All scientists, before they start any experiment, will attempt to "guess" the result, or make a hypothesis about the experiment. This shapes how and what they will do in their experiment. In this practical, before you start, you will write down your predictions of the results. It is best, if you can, to make your predictions quantitative - that is to say use numbers, including an explanation.

This process is very important in science. A good scientist can save a lot of time and money by using their experience and intuition to predict what they think will be useful experiments - instead of going down a dead end and getting nowhere.

IB Criteria Assessed

Criteria assessed	Level awarded
PE	/2
EX	/6
A	/6
EV	/6
C	/4

Apparatus:
An iron nail plus any other materials you may need.

Practice Exploration:
Design a procedure that will allow you to investigate the factor (or factors) that affect the strength of an electromagnet. This procedure should include the following sections:
- Defining the Problem and selecting variables:
- Controlling the Variables:
- Developing a procedure for collecting data:

Include a quantitative hypothesis for your investigation.

Analysis:
- Show the results for the experiment in a suitable table. Include uncertainties.
- Use suitable graphs to allow for an analysis to be carried out on your chosen variable(s)
- Include uncertainties in your processed data

Evaluation:
- Evaluate the data that you have collected and analyzed.
- Evaluate your own plan, including any modifications you had to make to overcome problems. Include an evaluation of the apparatus used.
- Suggest ways in which the procedure could be modified in order to improve it for the future.

INVESTIGATING CIRCULAR MOTION

Aim:
1. To investigate circular motion.
2. To use the formula for centripetal acceleration to calculate the mass of a rotating object.

IB Criteria Assessed

Criteria assessed	Level awarded
PE	/2
EX	/6
A	/6
EV	/6
C	/4

Apparatus:
Grass tube, rubber cork, string, meter rule, stopclock, masses and mass holder.

Theory:
When the system is in equilibrium $F_R = F_C$
(speed and radius of rotation remain constant)

Procedure:
1. Set the force, F_R, hanging from the string and record it.
2. Rotate the string and record the radius, r, and time period, T.
3. Change the force, F_R and repeat until 6 to 8 sets of data are recorded.

Diagram:

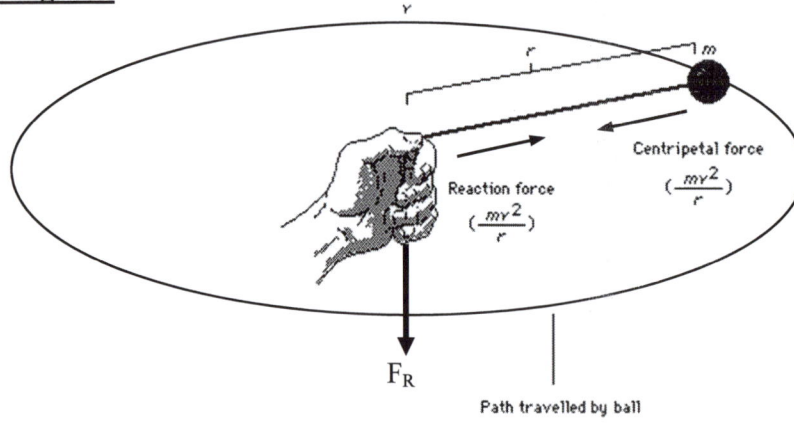

F_R = Reaction Force
m = mass of ball
v = speed
r = radius of rotation
T = time for 1 rotation

Analysis:
- In a suitable table record all of the data collected.
- Include actual and percentage uncertainties due to the measuring apparatus used.
- Draw a suitable graph that will allow you to use the formula for centripetal acceleration in order to calculate the mass of the rotating object. Include your error bars to indicate the uncertainty in the apparatus used. HL include a maximum and minimum gradient in order to determine the overall uncertainty in your calculated answer

Evaluation:
- Evaluate the results of the experiment - compare your results by taking the mass of the object using a balance.
- Evaluate the procedure, including any modifications you had to make to overcome problems. Include an evaluation of the apparatus used.
- Suggest ways in which the procedure could be modified in order to improve it for the future.

DETERMINING ENERGY DENSITY OF FUELS

Introduction:
All fuels that can be burned have a particular energy density, or how much energy is packed in per kilogram of fuel. Usually, for something like cars, we want fuels that have a high energy density, so we don't have to lug so much around with us. One benefit of gasoline over coal is that it has a higher energy density. In this lab, you will compare the energy density of fuels that can be used easily in the laboratory.

IB Criteria Assessed

Criteria assessed	Level awarded
PE	/2
EX	/6
A	/6
EV	/6
C	/4

Aim:
1. To compare the relative energy density of fuels.
2. To compare our measured energy densities with the actual literature values.

Theory:
The equation for energy density is....

> Energy density = energy of fuel / mass of fuel

A rough assumption will be that the energy from burning the fuel can be put directly into heating water.

If so, energy to heat water is = $mc\Delta T$, where

m = mass water
c = specific heat capacity of water
ΔT = rise in temp. of water

Diagram:

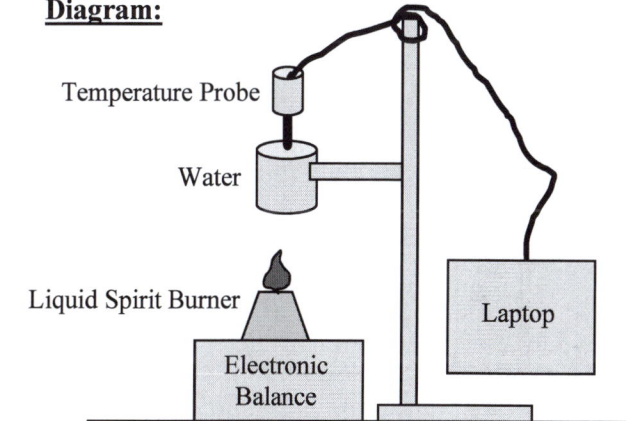

The mass of fuel used can be measured by the difference in mass of a spirit burner as it burns.

Procedure:
1. Set up the apparatus as shown in the diagram.
2. Use a reasonable amount of water given the size of your flame. 100 mL is a good starting point. Record this.
3. Record the initial temp of your water.
4. Lower the water (which should be on wire gauze) just over the flame.
5. Zero the scale so that it shows negative values as flame burns.
6. Set the laptop and temperature probe so that they measure water temperatures. Stir constantly.
7. Record appropriate data so as to obtain the energy density of water.
8. Record the uncertainties in the collected data.
9. Repeat this procedure using other fuels. Keep in mind that other fuels may heat the water much slower or faster, and you may want to use different amounts of water to allow for this.

Analysis:
- Collect appropriate data and present this in a way which will allow for easy interpretation.
- Include the uncertainties of measurement in your recorded data.
- Plot a suitable graph from which you can analyses the energy density of the fuels.
- Include the uncertainty in your calculated answer.
- Make valid conclusions related to fuel densities found. Compare the fuel densities relative to each other, and to literature values.

Evaluation:
- Evaluate the procedure, including any modifications you had to make to overcome problems
- Include an evaluation of the apparatus used.
- Suggest ways in which the procedure (and apparatus) could be modified in order to improve future investigations.

Original Lab Sheet by Ringo Dingrando

INVESTIGATING ENERGY

Aim:
So far in your physics course you have done 10 - 16 fairly short experiments. This is useful to give you practice in using equipment, solving practical problems, mathematical manipulation of results, understanding errors, taking accurate measurements, controlling variables, plotting graphs and testing laws. In the real world of science, scientists are paid to investigate a problem and provide answers. This investigation process may take weeks or even years. The next stage of your practical programme is therefore to apply the skills you now have in order to undertake a longer investigation. This will necessitate several stages :

IB Criteria Assessed

Criteria assessed	Level awarded
PE	/2
EX	/6
A	/6
EV	/6
C	/4

1. deciding on a topic that you would like to investigate.
2. planning a strategy for carrying out the investigation (to be approved by your teacher before you start).
3. collecting the necessary equipment.
4. carrying out the experiments you need to do.
5. doing the analysis of your results - plotting graphs, charts, etc.
6. evaluating your experimental results and reaching a conclusion.

Practice Exploration:
You are free to investigate any thing you like but the only conditions are:
- the investigation should involve 4 - 5 hours of lab work i.e. 3 double periods.
- the theme must be around the subject of "Energy".
- the project must be feasible within the limitations of time, equipment available and your capabilities.
- you must work alone and not in groups.

Here are some suggested topics if you are not able to think of your own :
- build and test the efficiency of a windmill.
- efficiency of a watermill.
- investigate efficiency of a solar cell.
- audit of energy use in the home or school.
- investigate efficiency of model aeroplane engine.
- energy output, cost and efficiency of a dry cell.
- compare efficiency of filament lamp and fluorescent light.
- efficiency of double glazing.
- effectiveness of insulation materials.
- compare efficiency of microwave and traditional oven.
- efficiency of bicycle and gears.
- energy analysis of toy train or car.
- efficiency of thermos flask.
- energy stored in a catapult.
- compare energy content of various fuels.
- human fitness.
- lifetime of torch bulbs.
- is a dry cell the most expensive way to buy electricity
- heat flow in a room due to a heater.
- how efficient are electric hand dryers ?

Investigating the Superposition of Sound Waves

Aim:
In your theory lessons you will have met this practical in which 2 speakers are connected to a signal generator and produce a line of constructive and destructive sound interference. Your task is to experiment with the conditions in order to find the optimum possible set - up that will produce the clearest pattern of loud and quiet. In the real world scientists spend a great deal of time "fine tuning" an experiment so that it will give the best possible results. It is this attention to detail that makes the best experimental scientists.

IB Criteria Assessed

Criteria assessed	Level awarded
PE	/2
EX	/6
A	/6
EV	/6
C	/4

Apparatus:
Signal generator, 2 identical speakers, metre rule, sound meter (if available)

Diagram:

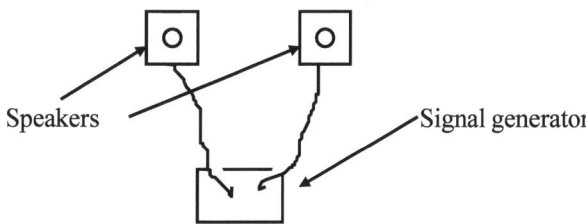

Procedure:
Vary the following conditions until you get the clearest possible pattern of loud and quiet sound as you walk along the line A - B (If you don't have a sound meter then you will have to use your ears).

1. distance between the 2 speakers.
2. loudness of the speakers.
3. frequency setting of the signal generator.
4. distance from the speakers to the line A - B.
5. does it work best in a large room or even outdoors?

Analysis:
- In your lab. report, list all the conditions you tried and the combination that you finally found to be best.

INVESTIGATING MAGNETS

Introduction:
You will probably have played around with magnets at one time or another. You may have noticed that a magnet will "stick" to some metals while it simply doesn't want to stick to others. You may have noticed that two magnets attract each other under certain circumstances while they push each other apart (repel) under a different arrangement.

Aim:
You are to investigate the force between two magnets

Apparatus:
Two small "magnadur" permanent magnets, plus any other materials you may need.

IB Criteria Assessed

Criteria assessed	Level awarded
PE	/2
EX	/6
A	/6
EV	/6
C	/4

Practice Exploration:
Design a procedure that will allow you to investigate the factor (or factors) that affect the force between two magnets. This procedure should include the following sections:
- Defining the Problem and selecting variables:
- Controlling the Variables:
- Developing a procedure for collecting data:

Include a quantitative hypothesis for your investigation.

Analysis:
- Show the results for the experiment in a suitable table. Include uncertainties.
- Use suitable graphs to allow for an analysis to be carried out on your chosen variable(s)

Evaluation:
- Evaluate the data that you have collected and analyzed. Make a suitable conclusion.
- Evaluate your own plan, including any modifications you had to make to overcome problems. Include an evaluation of the apparatus used.
- Suggest ways in which the procedure could be modified in order to improve it for the future.

INVESTIGATING LENZ'S LAW USING THE MOTION OF A FALLING MAGNET

Aim:
1. To find the motion of a magnet in free fall in a metal tube.
2. Use the data that you collected to Illustrate Lenz's and Faraday's Law.

IB Criteria Assessed

Criteria assessed	Level awarded
PE	/2
EX	/6
A	/6
EV	/6
C	/4

Diagram:

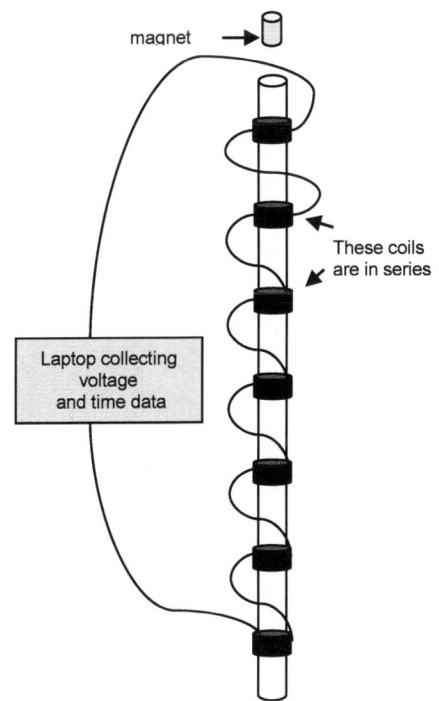

Procedure:

1. Obtain a long aluminum or copper tube, which has fixed coils attached to it.
2. Attach a wire from the top of the first coil to the Logger Pro interface box leads.
3. Run a second wire from the bottom of the first coil to the top of the second coil. Then run a wire from the bottom of the second coil to the top of the third coil. Continue this until you have all the coils connected in series.
4. Connect the bottom wire of the last coil to the 2nd lead from the Logger Pro.
5. Open the Logger Pro Software. Open the Experiments file. Open the voltage file.
6. Estimate the time that the magnet takes to fall through the tube. Add about 5 seconds to this and set this as your horizontal axis. Voltage will be your vertical axis.
7. Hit Collect, then drop the magnet in the pipe.
8. Measure the distance from the first coil to all the other coils
9. Construct a well labelled data table for you data.

Theory:
Lenz's Law and Faraday's Law both allow one to predict aspects of induced current when there is a changing magnetic field.

Analysis:
- Collect appropriate data.
- Plot a suitable graph from which you can analyses the motion of the magnet as it falls through the tube.

Evaluation:
- Using your graph, analyses the type of motion of the falling magnet.
- Using your knowledge of the laws of electromagnetic induction, explain why this type of motion occurs.
- Evaluate the procedure, including any modifications you had to make to overcome problems. Include an evaluation of the apparatus used.
- Suggest ways in which the procedure could be modified in order to improve it for the future.

INVESTIGATING ELECTROMAGNETIC INDUCTION

Aim:
From what we have learned through Faraday's law, we know that a changing magnetic field can induce a current in a loop of wire. Several factors can affect the strength of this induced current.

Apparatus:
- Wire
- Bar Magnets
- Galvanometer
- Anything else that could reasonably be found in a normal physics classroom.

IB Criteria Assessed

Criteria assessed	Level awarded
PE	/2
EX	/6
A	/6
EV	/6
C	/4

Practice Exploration:
Design a procedure to test how a certain variable (of your choice) may affect the strength of the induced current. This procedure should include the following sections.

- Defining the Problem and selecting variables:
- Controlling the Variables:
- Developing a procedure for collecting data:

Also include a hypothesis and a sketch graph of what you think will happen.

Analysis:
- Show the results for the experiment in a suitable table. Include uncertainties.
- Use suitable graphs to allow for an analysis to be carried out on your chosen variable(s)
- Include uncertainties in your processed data

Evaluation:
- Evaluate the data that you have collected and analyzed.
- Evaluate your own plan, including any modifications you had to make to overcome problems. Include an evaluation of the apparatus used.
- Suggest ways in which the procedure could be modified in order to improve it for the future.

INVESTIGATING THE EFFICIENCY OF A TRANSFORMER

Aim:
An engineer is often employed in order to test the efficiency of a machine under a variety of conditions. It could be that the machine is highly efficient over a narrow range but outside of this range, the efficiency drops drastically. It is important for a manufacturer or user to know what this range is. This is what you are going to do here. The aim is to produce an efficiency graph for a transformer.

IB Criteria Assessed

Criteria assessed	Level awarded
PE	/2
EX	/6
A	/6
EV	/6
C	/4

Procedure:
1. Connect up the circuit and set the power supply to 2 V a.c.
2. Measure the current and voltage on the primary and secondary sides of the transformer (you will have to disconnect and reconnect the meters as you only have 2).
3. Repeat the process for several different voltages up to 12 V.

Apparatus:
Demountable transformer, leads, 2 multimeters, variable a.c. power supply, 12 V light bulb.

Diagram:

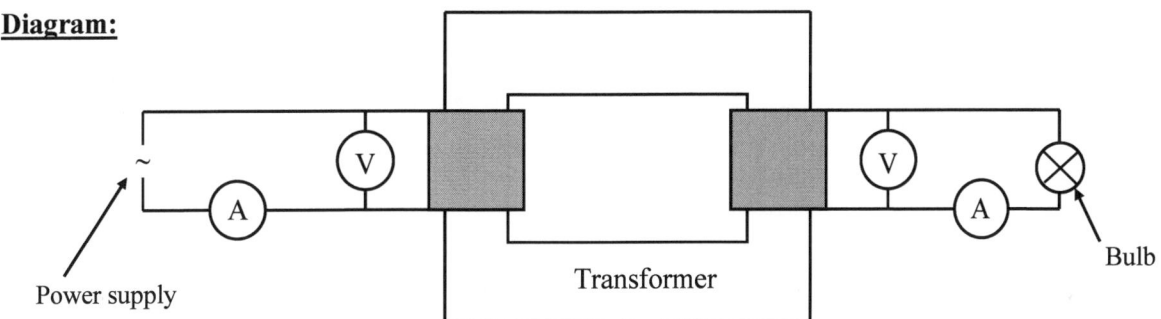

Analysis:
- Collect data for current and voltage on the primary and secondary side of the transformer.
- Present this data in a suitable table. Include uncertainties.
- Process your raw data in a way which will allow you to accurately calculate the efficiency
- Take into account any errors or uncertainties in your processed data.
- Plot a suitable graph which will allow a relationship between efficiency and voltage to be analyzed. Comment on this relationship

INVESTIGATING ENERGY TRANSFER AND ENERGY LOSS OF A ROLLING BALL

Aim:
In many situations in physics very useful results can be obtained by considering energy changes. Also in these changes, some energy is always "lost" and it is important to be aware of this and to attempt to explain what has happened to it. In this case you are going to examine a change from potential energy to kinetic energy when a ball rolls down a slope and from this find a value for "g" the acceleration due to gravity.

Also you will learn how to manipulate 2 equations so that a useful graph can be plotted from the resulting equation.

IB Criteria Assessed

Criteria assessed	Level awarded
PE	/2
EX	/6
A	/6
EV	/6
C	/4

Apparatus:
Ball bearing, stop clock, meter rule, ramp.

Diagram:

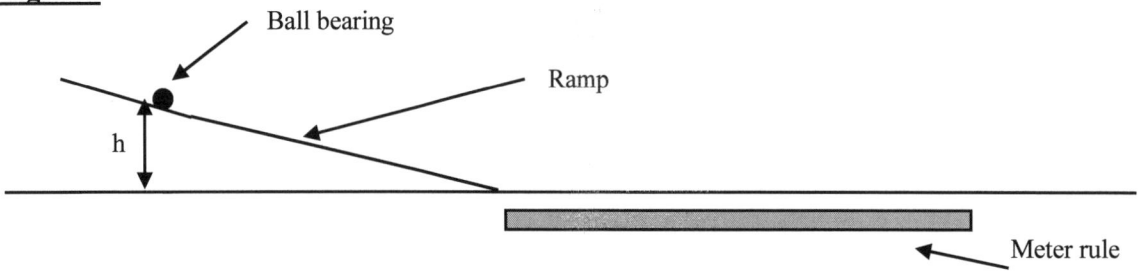

Procedure:
1. Release the ball bearing from a height "h"
2. Record the time taken for it to travel 1m once it leaves the ramp (or 2 m or 3 m if you think that this will be better).
3. Repeat this for several different heights.

Theory:

The equation for potential energy is   P.E. = mgh

The equation for kinetic energy is   K.E. = ½mv²

Analysis:
- Collect and record pairs of results for h and t including units and uncertainties.
- Present these data clearly.
- Combine the 2 equations above and from them plot a suitable graph from which you can calculate a value for "g".
- Include any errors or uncertainties in your processed data.
- Try to explain where energy has been "lost" and why this leads to a different value for "g".

INVESTIGATING THE FOCAL LENGTH OF A CONVERGING LENS

Aim:
To find the focal length of a converging lens by experimental method

Procedure:
1. Set the apparatus up as shown in the diagram, so that the lengths u and v are equal and so that the image of the candle appears in focus on the screen.
2. Record lengths u and v.
3. Move the candle away from the lens by 20cm more and adjust the screen distance (v), until the image is one again in focus. Record u and v in the table.
4. Repeat step 3, three more times (moving the candle further away from the lens). Record u and v each time.
5. Place the candle and screen back in the original position found in step 1.
6. Move the screen away from the lens by 20cm more and adjust the candle distance (u), until the image is one again in focus. Record u and v.

IB Criteria Assessed

Criteria assessed	Level awarded
PE	/2
EX	/6
A	/6
EV	/6
C	/4

Diagram:

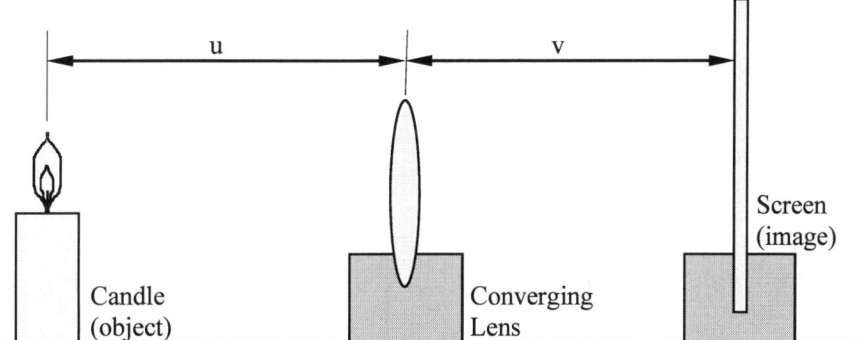

Theory:
The relationship between u, v and the focal length f for a converging lens is:- $\dfrac{1}{v}+\dfrac{1}{u}=\dfrac{1}{f}$

Analysis:
- Collect and record pairs of data (u and v) including units and uncertainties.
- Present these data clearly in a suitable table.
- Process your raw data in a way which will allow you to accurately calculate the value of f (the focal length of the lens)
- Take into account any errors or uncertainties in your processed data.
- Give a conclusion and explanation of your results; compare to quoted values if possible.
- Your explanation should include scale diagrams to explain some of the measurements taken. e.g.
    a. the object close to the lens (between pole and focal point)
    b. the object and image at the same position.
    c. the object far away from the lens.

Evaluation:
- Evaluate the above procedure and apparatus used, including limitations, weaknesses or errors.
- Suggest ways of improving the investigation.

Original Lab Sheet by Mike Dickinson

The Group 4 Project

The Group 4 Project
A practical project that MUST be completed by ALL IB Science students is the Group 4 Project. The Group 4 Project is an opportunity for students across the range of scientific disciplines to work together in order to investigate a topic from a range of angles. This interdisciplinary project allows Chemistry, Biology and Physics students (together with Computer Science, Sports Exercise & Health Science and Design Technology – if offered at your school) to engage in a collaborative project. Students taking Environmental Systems & Societies are not required to participate in the Group 4 Project. The emphasis of this project is on process rather than product and so, the Group 4 Project is not formally assessed.

Project Stages
The 10 hours allocated to the Group 4 Project, which are part of the teaching time set aside for developing the practical scheme of work, can be divided into three stages: planning, action and evaluation.

Planning
This stage is crucial to the whole exercise and should last about two hours.
- The planning stage could consist of a single session, or two or three shorter ones.
- This stage must involve all group 4 students meeting to "brainstorm" and discuss the central topic.
- The topic can be chosen by the students themselves or selected by the teachers.

A possible strategy is that students define specific tasks for themselves, either individually or as members of groups, and investigate various aspects of the chosen topic. At this stage, if the project is to be experimentally based, apparatus should be specified so that there is no delay in carrying out the action stage. Contact with other schools, if a joint venture has been agreed, is an important consideration at this time.

Action
This stage should last around six hours and may be carried out over one or two weeks in normal scheduled class time. Alternatively, a whole day could be set aside if, for example, the project involves fieldwork.
- Students should investigate the topic in mixed-subject groups or single-subject groups.
- There should be collaboration during the action stage; findings of investigations should be shared with other students within the mixed/single-subject group. During this stage, in any practically-based activity, it is important to pay attention to safety, ethical and environmental considerations.

Evaluation
The emphasis during this stage, for which two hours are probably necessary, is on students sharing their findings, both successes and failures, with other students. How this is achieved can be decided by the teachers, the students or jointly.
- One solution is to devote a morning, afternoon or evening to a symposium where all the students, as individuals or as groups, give brief presentations.
- Alternatively, the presentation could be more informal and take the form of a science fair where students circulate around displays summarizing the activities of each group.

The following pages outline one of the Group 4 Project that I undertook a few years ago while teaching at the United Nations International School, Hanoi, Vietnam. The overall theme of the project was "Science on and around the UNIS Campus" – students then investigated a wide variety of sub-topics within this general theme. Each of the small groups contained a physicist, a biologist and chemist (since UNIS did not offer the other 4 of the group 4 subjects) and while each student within the group was responsible for their particular science discipline, the whole group collaborated together to make these different aspects of physics to meld together into a coherent final report.

The following schedule and student sheets were put together into a booklet for each student and is just one way that the Group 4 project can be conducted. I happen to like this method of setting specific deadlines for the task and having the whole project performed over a two day period (ideally at the end of the first year of the IB course), but of course different teachers and different schools have chosen a variety of approaches to the timing of the project.

SCIENCE IB GROUP 4 PROJECT – INTRODUCTION

Guidelines
The Group 4 Project allows students to appreciate the environmental, social and ethical implications of science. It may also allow them to understand the limitations of scientific study. The emphasis is on interdisciplinary cooperation and the processes involved in scientific investigation. The exercise should be a collaborative experience where concepts and perceptions from across the science disciplines are shared. The intention is that students analyze a topic or a problem, which can be investigated in each of the science areas.

The project should take between 10 – 12 hours, and can be divided up into the following stages:

Planning	Identifying topics to investigate from the perspective of your science discipline(s) under the central theme of "Science at UNIS".
Definition of Activities	Deciding who is going to do what. Complete the planning sheet.
Action	Carry out the investigation, by either research or practical work. Share your findings with the other members of your group (and with students of other groups if appropriate). Make a PowerPoint presentation of your research.
Evaluate	Discuss with your peers the strengths and weaknesses of your project. Complete the evaluation form.

Note: Students taking two sciences must produce a presentation, which covers both sciences, and evaluate the project from the perspective of both sciences.

Timing

Thursday June 8th	1.55pm. Planning and Definition of Activities stages. IB Group 4 guidelines distributed to all students in order to familiarize themselves with what the project intends to do and achieve – what the IB is looking for. Students will be organized into groups. Group brainstorming session follows. Provisional title and outline decided on by all students.
Friday June 9th	8.10am. Collect completed planning sheet (including apparatus list) from all students.
Friday June 9th	8.10am – 3.05pm Action stage. Data collection and initial data analysis on and around the UNIS campus.
Wednesday June 14th	8.10am – 11.50am Action stage. Data collection and initial data analysis on and around the UNIS campus.
Wednesday June 14th	12.35pm. Presentation of work. All students are requested to prepare a PowerPoint presentation of their research. These presentations must be emailed to Mr. Dickinson by no later than 12 noon of Wednesday June 14th. **No presentation will be accepted at the start of the lesson.**

SCIENCE IB GROUP 4 PROJECT – QUESTIONS AND ANSWERS

1. *What is it?*
   Approximately 10 hours of research into an area decided by the students in that group and approved by a Science teacher.

2. *Does that mean that if I do two Sciences, I have to do two lots of 10 hours?*
   No, but you do have to investigate your area of research from the perspective of both of your science disciplines.

3. *How are the groups made up?*
   They are groups of 4 or 5 students from a mixture Science disciplines. For example, a group may have a Chemist, a Physicist, a Biologist and a couple of students from Sports Science, Computer Science or Design Technology. The groups are decided by the teachers, not the students!

4. *When does it take place?*
   During the last two weeks of grade 11. During the initial meeting, the groups will have to decide on their area of study. They will then have to assign tasks to each student that will take about 10 hours of research. These plans should be submitted on the form provided to the teacher that group is assigned to for approval and/or advice and modification. One and a half days will be put aside for students to do their research. This will take place on **Friday 9th June** from 8:10 a.m. until about 3:05 and **Wednesday 14th June** from 8:10 a.m. until about 11.50.

5. *What happens after that?*
   Once each team member has collected all the data, the team will have to meet to decide how they will organize their presentation. Each team will have to prepare a presentation of 10-15 minutes showing what they did and what they found out. This will take place on **Wednesday 14th June**, during the afternoon (periods 4 and 5).

6. *What form should the presentation take?*
   Each team should prepare a PowerPoint presentation, which will be a permanent record of their research. This may contain photographs, graphs and other data. Every member of the team must take part in the presentation. You will have to answers from the audience, which will include students and teachers.

7. *What will I have to hand in for assessment?*
   You need to hand in a report on your part of the project **for each Science studied** to the teacher of that subject. A good way to do this is to produce a miniature of your PowerPoint presentation (9 slides per page). Remember to include your planning and evaluation sheets. This report will form a vital part of your science practical scheme of work (4PSOW) so **don't lose anything**.

8. *What skills will be assessed?*
   The emphasis of this project is on process rather than product and so, the Group 4 Project is not formally assessed.

9. *What is the topic for research?*
   Science on and around the UNIS Hanoi, Tay Ho Campus.

10. *I'm doing Environmental Systems and Societies for my Group 3 subject. Do I need to take part in the Group 4 Project?*
    No – all ESS students are exempt from participation of the Group 4 Project.

Science IB Group 4 Project – Planning

Name .... Teacher's Name ....

Provisional title of project

....
....
....
....

Outline of project

....
....
....
....
....
....
....
....
....
....
....
....

Apparatus list

....
....
....
....
....
....
....
....

Group members and science disciplines

Physics ....
Biology ....
Chemistry ....

Science IB Group 4 Project 2006 – Planning Sheet (continued)

Group member	Data Collection Responsibilities (One set for each Science subject)
	……
	……
	……
	……
	……

Teacher Comments

……
……
……
……
……
……
……
……
……

Science IB Group 4 Project – Evaluation

Name .... Teacher's Name ....

Final title of project

....
....
....
....

Group members and science disciplines

Physics..
Biology..
Chemistry..

Evaluation

Strengths: ....
....
....
....
....
....

Weaknesses: ....
....
....
....
....
....

Summary of Conclusions

....
....
....
....
....
....
....

SCIENCE IB GROUP 4 PROJECT – SUMMARY

Each year, the grade 11 science students take part in a cross-discipline science project as part of the IB Diploma Practical Scheme of Work. Science is Group 4 on the IB Diploma hexagon and the "Group 4 Project" is an opportunity for students to work with areas of science that are often new to them.

Under the umbrella title "Science at UNIS", the 35 grade 11 students worked in groups of 4/5 investigating areas of the UNIS Tay Ho campus from the perspective of all 4 science disciplines. Each group comprised a Physicist, a Biologist and a Chemist.

The 2½ day (15 hour) project allowed for the collection and analysis of data from the four corners of the UNIS campus. Projects included:-

The Pond
- Oxygen and pH levels
- Algae content
- Are the fish happy?
- Natural vs. forced oxygenation
- Sunlight and shade

Air-conditioning
- How does it work?
- Effects on the body
- Environmental impact of CFCs released by the refrigerant
- Efficiency

Coffee
- Effects of caffeine on heart-rate
- Coffee growing regions of the world (with a focus on Vietnam)
- Paper vs. Styrofoam – avoiding MacDonald law-suits
- The half-life of caffeine in the body.

The Swimming Pool
- Safe levels of chlorination
- Disposing of chlorine safely
- What should the water temperature be for a comfortable swimming experience?

Litter
- Bacteria in the litter-bins
- Which rubbish can be burnt to generate electricity?

The Canteen
- Energy content in food
- Bacteria in the canteen
- Nutritional content of food Efficiency of microwave cooking

The students were enthusiastic and enjoyed the experience of taking part in a science project that that was fun without the pressure of extensive write-up and evaluation.

Appendix

POSSIBLE TOPICS FOR FURTHER INVESTIGATION

While on my Level 1 workshop in Cali, Colombia 1997, Brian Seve gave me this list of titles for potential investigations – either to be added to the 4/PSOW, for possible ideas on topics or subtopics for the school's Group 4 Project or perhaps when a student is interested in doing their Extended Essay in Physics but can't think of a theme or a title.

- The shutter speeds of a camera
- The accuracy of aim of an air rifle, catapult, or improvised gun
- The true path of a ball thrown in air
- Water drops failing on water (flash pictures?)
- Splashing of moving drops hitting solids
- A narrow water trough as an accelerometer
- The profile of a rotating water surface
- The precession of a gyroscope
- Comparisons of human reaction times (between individuals; for different stimuli)
- Time taken by a switch to make or break contact
- Bouncing of relay contacts
- How much does the air pressure in a football matter?
- The performance of a firework rocket
- The bounce-time of a ball
- Factors affecting the friction of steel on ire
- The effect of oil films between sliding metal surfaces
- Does water absorb ultra-violet light?
- How long does the flash from a flash bulb last?
- How long does the flash from a xenon stroboscope last?
- How does the light coming through a slotted wheel stroboscope vary with time?
- Study the motion of a ball rolling on a turntable
- What does an air track collision look like from a moving point of view? (Moving camera)
- The distribution of speed, or of energy, among balls rolling randomly in a shaking tray
- The possible orbits of a pendulum bob
- The motion of the lip of a vibrating wire
- The performance of a water pump
- The performance of a fan
- The thrust of a propeller (in air, or in water)
- The energy delivered by a catapult
- Load and speed variations of a model aero-engine
- The fuel consumption of a model aero-engine
- The temperature changes and cooling of a model aero-engine
- The air supply to a model aero-engine
- Reduction of noise from a model aero-engine
- Factors affecting beam bending
- Factors affecting the buckling of a beam under compression
- Factors affecting the flexing of a rotating shaft
- The strength of girders of different construction (use balsa wood)
- The energy stored in a spiral clock spring
- Factors affecting the design of a good paddle wheel
- Making strong concrete bars
- The fracture of concrete by impact forces
- Effects of reinforcement on concrete
- The strength of fiberglass repairs (commercial fiberglass kits)
- Ice is said to be made less brittle by freezing sawdust into it. Is it?
- Variation of flow behavior with rate of strain (silicone putty)
- Effects of heat-treating razor blades
- Heat-treatment of steel
- Heat-treatment of copper
- Heat-treatment of glass
- Flow patterns in glycerin (see Shapiro, A. H. Science Study Series, Shape and flow, Heinemann)
- Perspex is said to 'remember' that it has been deformed, for a while. Does it?
- The strength of human hair
- The strength of paper
- The properties of glued joints
- Making long lasting soap films
- Adhesion of glues to metals, fabrics, etc.
- How finely woven must umbrella material be?
- The changes in melting point of a solder with composition
- What is necessary for solder to flow?
- The strength of a soldered joint
- The bouncing of steel bails on glass
- Impact cracks when steel balls are dropped an glass
- Dents made in metals by balls pressed on them (Brinell hardness test)
- The heating and cooling of stretched rubber
- The creep of stretched rubber
- The strength and fracture of taut rubber bands
- The effect of temperature on stretched rubber
- Changes of length of hair with moisture content
- Factors affecting the growth of crystals
- The sagging of taut wires loaded in the middle
- The shape of a suspended loose chain
- Will a hole at the end of a crack help to stop the crack from spreading?

From a list compiled by Brian Seve

- What factors influence the production of good, uniform bubble rafts?
- The effect of various sorts of perforations on tearing paper
- The pressure-volume relation for a rubber balloon
- The effect of temperature changes on the flow of motor oils
- The design of a flow meter
- Reduction in pressure with fast flow (Bernoulli effect)
- Calibration of a V-slot flow meter (rate of flow from height of water in a V shaped slot)
- The drag on spheres and other shapes in an air stream
- The resistance to water flow of various plumbers' fittings (pipe, bends, etc.)
- The drag on objects towed in water (changes with length, depth of water, and many other factors)
- When does water flow in a tube become turbulent?
- The effect of changing the size or shape of the wings of a glider
- The penetration of projectiles into soft materials
- Load and speed variations for a parachute
- A water-driven rocket
- Measuring the viscosity of air
- Factors affecting the performance of an air track vehicle
- Making very big drops (oil in water and alcohol mixtures)
- How do Plateau spherules form?
- Soap films formed on spirals and other wire shapes
- The behavior of bubbles rising in liquids
- The noise made by a kettle just before it boils (singing)
- The airflow in a room with a heater
- Smoke rings (a box with a hole at one end, and a flexible diaphragm at the other)
- Vortex rings in water (drop colored water drops onto clear water)
- How much can a container be overfilled with water?
- How does water drip from a narrow jet?
- Variations in damping of a pendulum in air
- Water from a tap running into a flat basin sometimes forms a smooth ring of water,
- with a circular edge beyond which the flow is rougher. What decides the size of the ring?
- Where does dust collect? Why?
- Stiff standing rods will oscillate in an airflow. Investigate.
- The supporting of a ball on a jet of air, or of water
- The behavior of coupled oscillators
- How much damping is needed to stop oscillations?
- Variable damping of a galvanometer
- Oscillations of drops
- Oscillations of rubber sheets
- Oscillations of soap films
- Oscillations of metal discs
- Oscillations of thin panels (e.g. doors, sheets of hardboard, sheets of metal)
- Oscillations of wire rings
- Oscillations of solid bars (notes from a xylophone)
- The factors affecting the performance of a sensitive flame
- How long does a sound last in a large hall?
- The propagation of sound at low pressures
- Can the motion of air in a sound wave be made visible?
- Slopping modes of oscillation in tanks of water
- How to isolate laboratory apparatus from vibrations
- 'Pearls in air'- what makes them form easily? (See Nuffield O-level Physics, Guide to experiments IV, experiment 21b)
- The resonance of a 'ticket timer'
- The frequency characteristics of a cheap gramophone pick-up
- The frequency response of a one-transistor amplifier with feedback
- Photographing waves on strings or springs
- The wakes of boats
- Waves in moving water
- Speed of waves in shallow water
- Breaking of waves
- The speed of ripples on water
- What are the shadows of waves on a ripple tank shadows of?
- The directional properties of a television aerial
- Variation in response of a dipole with length of the dipole
- Frequency range of a microphone
- Audible range of humans and animals
- The diffraction of sound waves
- Producing and detecting ultrasonic waves
- The pressure changes in the sound from an explosion
- Reflection or absorption properties of materials for microwaves
- Reflection or absorption properties of materials for sound waves
- Sound-absorbing tiles sometimes have perforated hardboard over an absorbent layer. Does the hole size matter?
- The behavior of a loudspeaker cabinet at low frequencies
- The penetration of sound through double glazed panels
- Waves in circular dishes
- How good is a wax lens for microwaves?
- The colors of thin films of oil on water
- 'Shadows' of hot air from flames or heated objects
- The field of view of a simple telescope
- The depth of focus of a simple telescope
- The depth of focus of a microscope
- The resolution of a microscope

Possible Topics for Investigation

- Depth of focus of a camera
- Photography through a microscope
- Patterns in stressed materials between crossed polaroids
- Moiré fringes (Patterns from overlapped regular grids)
- Detection of small motions by interference methods (thermal expansion, compressibility)
- How much light is reflected at various angles by glass?
- The sensitivity of Kodak PI 53 paper at various wavelengths
- The adaptation to dark of the human eye
- The resolution of close-spaced objects by the eye
- Does photographic film fog equally if the light is bright and the exposure short, or if the light is dim and the exposure long?
- How big are the grains in a photograph?
- How fast must a flicker be before it stops being observable?
- Make a diffraction grating by photographic reduction, and test it
- Do people vary in the range of wavelengths they can see?
- How quickly does the iris of the eye contract when the light is made brighter?
- Does the resolution of the eye depend on the illumination?
- The performance of a pin hole camera
- How much is scattered light polarized?
- A dynamo as a speedometer (conversion to accelerometer?)
- The efficiency of a dynamo
- The efficiency of an electric motor
- Load and speed variations of an electric motor
- Efficiency of a transformer
- Saturation effects in a transformer
- Effect of air gaps in transformers or electromagnets
- Eddy current losses in transformers (solid core)
- Stray fields around transformers
- The time taken for a fuse to blow
- The conduction of electricity by pencil lines on paper
- Conducting paper as a model for electric potential variations
- Potential variations in a tank of conducting liquid
- The time taken for ions to recombine (e.g. blown down-wind of a flame)
- How good are 10 per cent radio resistors?
- How good are 20 per cent radio capacitors?
- Torque-speed variations of a gramophone motor
- Energy emitted by a lamp bulb
- Lifetime of torch bulbs
- Does a photo-transistor respond instantly?
- Variations of resistance with strain
- How sensitive can a Wheatstone bridge be made?
- Resistance changes of human beings with variations in emotional state
- The running down and recovery of a dry cell
- How much charge can a home-made accumulator store?
- Electrolytic capacitors are said not to lose all their charge if short-circuited after being charged for some time. Is it so?
- An electroscope as a voltmeter
- The sensitivity of an electroscope as a charge measuring device Moving coil
- Milliammeters as ballistic galvanometers
- Make a capacitor microphone
- The variation of the field of a small coil with angle
- The contraction of a spiral carrying a current
- The effect of thickness of metal on eddy current forces
- How high will a 'jumping ring' jump? (A ring over an iron core with a coil carrying a.c. on the core)
- Frequency dependence of the impedance of an iron-cored inductor
- The dependence of the speed of a d.c. motor on field current
- Change in length of a nickel rod in a magnetic field
- The voltage from a thermocouple
- Temperature variations of transistor currents
- Is it true that a dry cell is the most expensive way to buy electricity?
- The design of an alternating current ammeter
- Behavior of two LC circuits coupled together
- The design of an electronic exposure timer
- The energy balance of a photocell
- Electrical noise in a hot resistor
- Does a flame conduct electricity?
- Does hot air conduct electricity?
- What factors make for good deposits of copper in electrolysis?
- What factors affect heating by eddy currents?
- How does the resistance between two points on a conducting sheet vary with distance?
- How does the resistance between two flat plates in a tank of conducting liquid vary with their spacing?
- Make an electrostatic dust collector
- Magnesium oxide smoke collects in long fibers on electrodes at high potentials. Investigate. (Exclude draughts.)
- How does the resistance in an LC circuit affect the resonance?
- How does the electron current in a radio valve vary with filament temperature

From a list compiled by Brian Seve

Printed in Great Britain
by Amazon.co.uk, Ltd.,
Marston Gate.